Classic old-growth forest provides a stunning backdrop for fishermen on Baker Lake in northwestern Washington.

FRAGILE MAJESTY

The Battle for North America's Last Great Forest

KEITH ERVIN

The Mountaineers • Seattle

The Mountaineers: Organized 1906
" . . . to explore, study, preserve, and enjoy
the natural beauty of the Northwest."

Published by The Mountaineers
306 Second Avenue West, Seattle, Washington 98119

Published simultaneously in Canada by Douglas & McIntyre, Ltd.,
1615 Venables Street, Vancouver, B.C. V5L 2H1

3 2 1 0 9
5 4 3 2 1

Manufactured in the United States of America
Edited by Jim Jensen
Drawings by Linda Wilkinson
Cover photograph by Bob and Ira Spring
All text photos by the author unless otherwise noted
Frontispiece: Classic old-growth forest provides a stunning backdrop for fishermen
on Baker Lake in Northwestern Washington.

Library of Congress Cataloging in Publication Data

Ervin, Keith.
Fragile majesty: the battle for North America's last great forest
/ Keith Ervin.
p. cm.
Includes bibliographical references.
ISBN 0-89886-230-2. -- ISBN 0-89886-204-3 (pbk.)
1. Old growth forests--Northwest, Pacific. 2. Forest
conservation--Northwest, Pacific. 3. Forest Ecology--Northwest,
Pacific. 4. Logging--Northwest, Pacific. I. Title.
SD387.043E78 1989   89-38414
333.75'16'09795--dc20   CIP

Acknowledgments

It would be impossible, in one page, to recognize all those who assisted me in my work on this book. I wish to acknowledge a few who were especially generous in sharing their time and their expertise. The conclusions I have reached are strictly my own.

I am beholden to the researchers who explained the intricacies of forest ecology and the art of growing trees, the hard-working folks who have devoted their careers to producing timber or managing the public lands, and those struggling to save the great forest. Thanks to Eric Forsman, Jerry Franklin, Len Ruggiero, Chris Maser, Harriet Allen, Eric Cummins, Ann Potter, Dale Cole, Elliott Norse, Jeff Cederholm, Rex McCullough, Cheryl Talbert, Walt Robinson, Jr., Gary Jones, Dewey Bryson, Ron Smith, Dick Whitmore, Mike Roberts, Gus Kuehne, Dick Pierson, Tom Ambrose, Bob Dick, Ron DeHart, Doug MacWilliams, Barney Dowdle, Richard Haynes, Mark Houser, Dave Larson, Rob Harper, Tim Boyd, Jean Durning, Brock Evans, Bill Arthur, Pam Crocker-Davis, Chuck Sisco, Charlie Raines, Andy Kerr, Rick Brown, Andy Stahl, Frank Ancock, Mitch Friedman, George Draffan, Greg Wingard, Kris Fleming, Gino Lucchetti, Kurt Russo, Dave and Juanita Jefferson, Jewell James, and Bill James.

For the use of photographs, I thank Weyerhaeuser Company, Jeff Hughes, Adrian Dorst, the Tulalip Tribes, and Tracy L. Fleming. The late Stu Bledsoe opened my eyes to the loss of prime timberland to urban sprawl. To Tom Campion, my gratitude for a top-notch introduction to the old-growth forest.

The Bullitt Foundation, Washington Environmental Foundation, and Mountaineers Foundation provided the most tangible kind of encouragement available: money. Special thanks to Bev Dahlin, Donna DeShazo, Steve Whitney, Marge Mueller and Rick May of The Mountaineers, and to Linda Wilkinson for her excellent artwork.

By far the greatest support was that of my wife, Cindy. I could not have written this book without her encouragement and extraordinary sacrifices. She managed the household, produced the family income, and gave birth to Sarah. To Cindy, my love and my deepest thanks.

Contents

1

A Life In Our Hands

Its nose twitching, the small mouse makes its way along the narrow branch. The animal's quivering body speaks of caution, yet it remains oblivious to its impending doom.

The owl is perched on a higher branch, his sharp eyes riveted on the prey. Silently, motionlessly, the bird calculates the best method of attack. Suddenly, he swoops down and, with scarcely a pause in his wingbeats, snatches the rodent in his talons. A few thrusts of his powerful wings and the deadly hunter is safely roosting in another tree, savoring his meal.

Watching the northern spotted owl are three other sets of eyes, the eyes of men. Two biologists and I have walked to this patch of woods on the northern edge of Washington State's Olympic Peninsula, in pursuit of the most controversial bird in America. For the spotted owl has become both a powerful symbol and the chief focus in the battle now raging to save North America's last great forest: the "old growth" of the Pacific Northwest.

On our way into the forest, before we heard the owl's first *Hoo-hoo!*. . . *Hooooooo!*, one of my companions, Stan Sovern, spotted a logging tower in a clearcut on the ridgetop. When Sovern climbed the ridge earlier in the year to band a female owl, the clearcut wasn't there. Sovern and Eric Forsman are after the male today. Their search will be punctuated by the occasional whine of a chain saw and the toot of a whistle on the logging site.

The owl, ready for another mouse, flies to a perch closer to the men. Forsman, the lanky young biologist directing this research sponsored by the U.S. Forest Service, walks slowly toward the bird, carrying a fishing rod with a nooselike loop at the tip. The bird calmly sits on the branch as Forsman slowly raises the pole, slips the loop around its neck, then gently lowers the now-struggling owl to the ground. Sovern takes the owl in his hands and holds the bird's wings tightly as Forsman clips colored metal bands to both legs. The bands will help scientists in their studies of the spotted owl population on the peninsula.

Forsman and Sovern carefully examine each feather on both wings to determine the progress of molting. Then they gently stuff the bird into a bag and weigh him. The men freeze, then sigh, in response to the earthshaking crash of an old-growth tree. The ancient forest is falling.

The biologists release the bird and "mouse" him once again. His second

9

meal in his mouth, the bird watches us from a branch far beyond the reach of Forsman's fishing pole. Finally, tired of these games with humans, the owl flies off.

Walking down the hill, we notice what we had missed in our haste on the way up. A series of yellow markers are stapled to tree trunks. "BOUNDARY CLEAR-CUT," they read. The signs are familiar to those who roam the national forests of the Pacific Northwest. Part of the owl's home was torn down today. The trees in which we saw him roosting will fall soon.

The spotted owl researchers don't mix much with townsfolk here in logging country. Except when they're in the field, they stick close to the Forest Service work station where they sleep in a mobile home and raise mice for trapping owls. Likewise, the Washington Wildlife Department biologists surveying spotted owls on state land don't often leave their cabin to spend time in the nearby town of Forks.

Already, the findings of Forsman and other researchers have inflamed local passions. Logging is by far the biggest industry on the Olympic Peninsula. The nearby town of Forks bills itself, plausibly, as the logging capital of the world. Spotted owl researchers are about as popular here as freedom riders were in the Deep South. The café in Forks sells T-shirts and logging suspenders that read, "SPOTTED OWL HUNTER." A popular float in Forks' Fourth of July parade featured a logger chasing a spotted owl with a chain saw. Local managers of the Washington Department of Natural Resources--stewards of the state forestlands--drew laughs when they handed out bogus spotted owl hunting regulations.

The northern spotted owl, *Strix occidentalis caurina*, is a subspecies that has become symbolic of the battle over the last of this continent's virgin forests. The bird is unusual, though perhaps not unique, in its specialization. It doesn't just require forest for its survival. To live and breed at viable levels, it needs a special kind of forest. Only a conifer forest will do. Only a forest in the temperate zone of the Pacific Coast between northern California and southern British Columbia. Only a forest essentially undisturbed by humans. Preferably a forest in which the oldest trees took root anywhere from 200 to 1,000 years ago. And large enough that a breeding pair can forage over hundreds or even thousands of acres.

The spotted owl has become a symbol of the fragile ecosystem unique to the old-growth forests of the Pacific Northwest. It has become a symbol, too, of the threat that loggers and their families feel in the face of a growing movement to save the ancient forests. "They're putting people out of work for a danged bird," fumes one logger. "That's just sheer stupidity." Leaders of the forest-products industry have taken to calling the spotted owl "the billion-dollar bird."

Though attention has focused on the owl as an emblem for the forests, it's only one part of a greater issue. An entire ecosystem, one of the grandest on the planet, is at stake. This ecosystem is home to a yet-uncatalogued range of plant and animal life. The majestic woods are mostly gone. Less than one-fifth of the old growth that once covered the landscape of western Oregon and western Washington still stands. No one has tallied the full dimensions of this loss because no one has made a comprehensive inventory of the remaining old growth, much less figured out how much has already been cut down.

What remains is going fast. In the national forests alone, 48,000 acres of virgin forests are being cleared to feed the sawmills and pulp mills of Oregon and Washington each year. The U.S. Bureau of Land Management (BLM) is selling off another 22,000 acres. Within fifteen years--barely a summer's afternoon in the time frame of the forest--the last of the old growth on state and private lands will be gone.

Then there will be only what's left in the national parks, national forests, and, in Oregon, on land administered by the BLM. To the timber industry, that's a lot of land. Over a million acres of older forests on federal land is being preserved in the Douglas fir region west of the Cascade Range crest for a number of reasons: wilderness and recreation values, habitat for the spotted owl and other species, or site-specific engineering problems such as steep and unstable soils or regeneration difficulties. A million acres that could be providing jobs and profits. A million acres that timber industry lobbyists claim the spotted owl doesn't even need.

Compared to what once was, a million acres of protected forest is a pittance. When European settlement began, an estimated 850 million acres of what are now the lower forty-eight states were covered by virgin forests. The trees were cut or burned down, first to make way for the settlers, then to produce lumber. The forests of the Pacific Coast and the Rockies are all that's left to give us an inkling of what this land may have been like three centuries ago. Even in the Northwest--where settlement began in earnest only 150 years ago--the lowland forests where trees grew to almost unimaginable size in rich alluvial soils have long disappeared.

What's left is being fragmented by patchwork clearcuts into isolated stands that are losing their ability to support spotted owls and other creatures of the ancient forests. The remaining old growth, mostly at higher elevations and on steep hillsides, sometimes is described as "the dregs."

Ah, but what dregs! In biomass alone, the Pacific forests are rivaled by no other forests on earth. The rain forests of the tropics are small things by comparison. Below the Northwest's mantle of tall trees is a unique world of plant and animal life. The Olympic rain forest's rich mantle of mosses, lichens, club mosses, and liverworts is thicker and heavier than that of any other temperate forest in the world. The Northwest forests have been called, justifiably, "the world center of mushroom diversity." Beyond the spotted owl, animals as diverse as bald eagle, red tree vole, and rough-skinned newt find their best habitat in ancient forests. Only now, when this ecosystem is

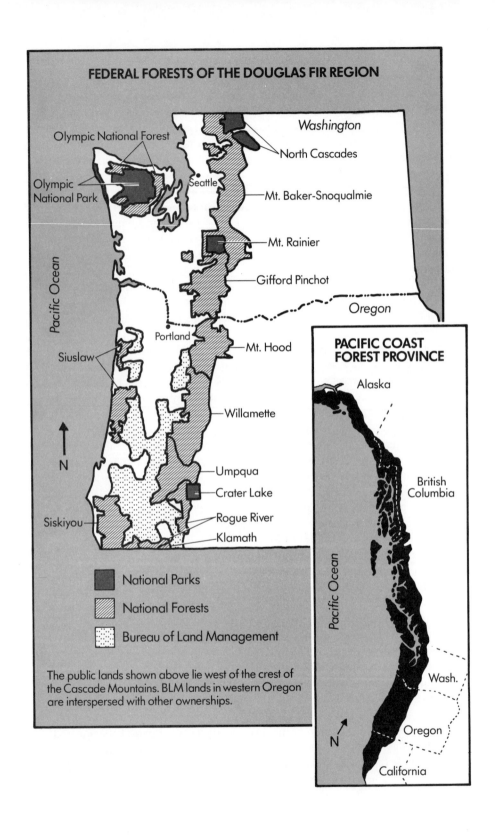

FEDERAL FORESTS OF THE DOUGLAS FIR REGION

Olympic National Forest

Washington

North Cascades

Olympic National Park

Seattle

Mt. Baker-Snoqualmie

Pacific Ocean

Mt. Rainier

Gifford Pinchot

Oregon

N

Portland

Mt. Hood

Siuslaw

Willamette

Umpqua

Crater Lake

Rogue River

Siskiyou

Klamath

National Parks

National Forests

Bureau of Land Management

The public lands shown above lie west of the crest of the Cascade Mountains. BLM lands in western Oregon are interspersed with other ownerships.

PACIFIC COAST FOREST PROVINCE

Alaska

Pacific Ocean

British Columbia

N

Wash.

Oregon

California

in danger, have scientists come to appreciate the magnificence of these forests.

No one has done more than Jerry Franklin to demonstrate the biological opulence and uniqueness of the Pacific Northwest's old-growth forests. The Forest Service's Pacific Northwest Research Station in Corvallis, Oregon, calls Franklin its chief plant ecologist. The University of Washington's College of Forest Resources calls him professor of ecosystem analysis. Some people simply call him the guru of old growth.

Franklin grew up playing in the second-growth forests near his Camas, Washington, home. Even more special to him than those woods were the old-growth groves of the Gifford Pinchot National Forest, where his family spent vacations camping. Forests were in the family line. Franklin's parents even gave Jerry the fitting middle name of Forest. The mystique of the forest never wore out. Even today, the scientist speaks of the "aura" of old growth and of the "inspiration" he draws from the ancient forests.

The son of a worker in Crown Zellerbach's Camas paper mill, Franklin is no wild-eyed environmental radical. His soft voice and avuncular, down-home manner have only boosted his credibility in the highly politicized atmosphere surrounding the debate over management of the last old-growth forests.

Franklin has been studying the forests of Washington and Oregon since the late 1950s, when he was a forestry graduate student at Oregon State University. It wasn't until 1970, though, that serious research into the old-growth ecosystem was first undertaken. As deputy director of the Coniferous Forest Biome research project, Franklin played a crucial role in obtaining funding from the National Science Foundation and lining up researchers from Oregon State and the University of Washington.

"I think the bottom line is we learned the old-growth forest was distinctive in a number of its characteristics from younger forests," Franklin recalls, leaning back from the desk in his university office. "It's not just a younger forest grown up to a larger size. It performs some functions very well, and it has a different kind of structure than a younger forest does and because of that provides habitat for a different set of animals. It was interesting because the Forest Service had stopped doing research on old growth in about 1960 because they felt we had learned everything we needed to know about it--which was basically how to cut it down and regenerate a young forest."

Initially, the researchers weren't investigating old growth per se, rather the coniferous forests of the Douglas fir region. The focus slowly shifted toward ancient forests as it became apparent that the most distinctive features of the Northwest woods were precisely those that took centuries to develop. It was the spectacular biomass and vegetative richness of old growth that stood out in study after study. The biomass accumulated by the big trees in old growth, it turned out, produced a unique set of flora and fauna. By

1981, scientists knew enough about old growth to describe it in a landmark report authored by Franklin and seven associates, *Ecological Characteristics of Old-Growth Douglas-Fir Forests.*

As Franklin explains, big trees are the engine that drives the old-growth ecosystem: "The trees are large, old, the crown structures are very complex, the dead wood component--standing dead trees and logs--is very conspicuous and that's in part because many of the species are quite decay-resistant so that these large woody structures disappear only slowly. And because of the canopies, they modify the environment within the forest incredibly so that the moisture and temperature conditions are totally different than they are out-side or even in young forests. And that's one reason why many organisms find it to be a very favorable environment. It's extremely stable, the extremes are highly muted."

In the microclimate of the old-growth forest, animals and plants find warmth and shelter from the snow in winter. They find coolness and moisture in summer. From fog and clouds, the crowns of tall trees wring moisture--in some drainages accounting for one-fourth of total precipitation in some watersheds. The wind is still on the forest floor, rarely blowing harder than two miles an hour. The irregular canopy of old growth lets in enough light to support a far richer understory than is found in young forests.

The old-growth ecosystem begins with big trees, both live and dead. Another ecologist, Elliott Norse of The Wilderness Society, says of the forest giants, "These trees are as exceptional in the plant world as whales are in the animal world. Ancient conifers are the whales of the forests." The Pacific Northwest forest range, from northern California to southeast Alaska, produces the biggest conifers in the world. Only the huge eucalyptus-dominated forests of Australia and New Zealand come close to the biomass of the Northwest old growth. On average, the Douglas fir and noble fir forests of Oregon and Washington contain three times the biomass of tropical rain forests. This "huge photosynthetic factory," as Jerry Franklin sometimes calls old growth, accumulates biological mass more efficiently than any ecosystem on earth. Trees simply grow crazy in the Northwest.

Temperate forests typically are deciduous or mixed deciduous-conifer stands. The Pacific Coast, with its distinctive weather patterns, breaks the mold. Conifers overshadow deciduous trees a thousand to one by timber volume. Although there is ample precipitation for trees, rain and snow fall primarily during the winter months. Drought is an annual summer event. Seattle typically receives three inches of rain between June and August. Less than half that amount falls on Medford, Oregon. Deciduous trees can carry on photosynthesis only during the warm months when their leaves are out. If rainfall is inadequate during that critical time, deciduous trees just can't make it. Conifers, able to produce carbohydrates year-round, have a tremendous competitive advantage on the West Coast.

Conifers aren't just unusually plentiful here, they grow like nowhere else. Ten genera of conifers grow in the Northwest; the largest and longest-living species of each is found on the coast. "They have a genetic makeup that simply enables them to persist and grow for very long periods of time," Franklin observes. "Whereas a loblolly pine in the Southeast is pretty much pooping out by the time it gets to be fifty or sixty years old, a Douglas fir is only beginning to get started at that age."

Taken individually, none of the attributes of old growth reported in *Ecological Characteristics* was terribly surprising. Of course an ancient forest is dominated by big trees. Of course standing dead trees are abundant, as are large logs on the ground and in streams. Anyone who looked could see that mushrooms, mosses, lichens, and liverworts grew in profusion. The report galvanized the scientific community, and began to ripple through the national forests' interest groups, because it showed for the first time how the whole system worked and how it differed from other forests.

Franklin and his colleagues pointed out that the deep, irregular crown of an old-growth Douglas fir provides "ideal habitat" for such specialized creatures as the red tree vole, northern flying squirrel, and northern spotted owl. The old-growth canopy provides a home for an estimated 1,500 species of invertebrates. Large logs offer animal habitat and seedbeds for young trees. Nitrogen, often in short supply in forest soils, is built up by lichens in the forest canopy and by bacteria that proliferate in rotting logs. Large logs also provide a home for mammals that spread underground fungi. The food chain, or "energy cycle," of small old-growth streams begins with woody debris rather than green plants. Few plant or animal species are found *only* in old growth, but many find their best habitat there, and some may require a reservoir of old growth for their survival.

Between 175 and 250 years typically are required for old-growth characteristics to develop under natural conditions. Old growth "begins to come into its prime" after 350 years, says biologist Andy Carey. The term "old growth" has been used by foresters for many decades. Yet it wasn't until the mid-1980s that scientists even tried to develop an ecological definition of the term. Before that, foresters and researchers used whatever definition they found most convenient. Those wildly varying definitions generally were based on a single criterion such as tree size or age. Some foresters used the term old growth for anything that had not been cut. In 1986, the Forest Service's Old-Growth Definition Task Group, chaired by Jerry Franklin, proposed a definition that could be applied to a stand by a forester with a tape measure. The definition was both ecological and quantifiable. There had to be a certain mix of species, a certain number of live trees of various sizes per acre, along with with a specified number of snags and logs of defined size.

The ecological definition set objective standards by which anyone can determine whether a forest is to be considered old growth. The standards were somewhat arbitrary; there's no magic point at which a forest is suddenly transformed from a "mature" forest into "old growth." Jerry Franklin speaks of "degrees of old-growthedness" and a "continual gradient of old-growth characteristics. . . It becomes a little more sophisticated than a simple yes-or-no answer, 'It is or it ain't.'"

The forest in which Eric Forsman and Stan Sovern captured the spotted owl wouldn't meet the ecological definition of old growth. It had the mix of species and the deep, multilayered canopy characteristic of old growth, but it lacked the requisite number of centuries-old trees to meet the definition. A forest in the process of becoming old growth, it supports a wider range of wildlife than an even-aged tree farm but probably fewer species than true old growth. By 1989, Franklin was urging scientists to supplement the old-growth definition with more flexible measures of a forest's ecological structure.

The either-or definition is the closest thing that exists to a scientific consensus on the minimum standards for old growth. Yet even after the definition was published, national forests continued to release draft management plans that bore no relation to the definition. By continuing to use nonecological definitions, the Forest Service was able to say that millions more acres of old growth remained on its land than the ecological approach indicated.

The scientists who wrote *Ecological Characteristics* warned that the ancient forests were rapidly disappearing. It was true that the Forest Service would be selling old-growth timber for the next forty years and that some of the virgin forest was protected in national parks, wilderness areas, and research natural areas. "Nevertheless," the scientific group reported, "these reserves occupy less than 5 percent of the original landscape, and the end of the unreserved old-growth forests is in sight."

If an old-growth forest were a stage, the principal players would be the big trees. Like the protagonists of a Shakespearean drama, these nobles shape the world in which a host of lesser trees and plants live. Those others play supporting roles. Just as great men and women do much to shape society, so these trees give an ancient forest its structure.

From southern British Columbia to northernmost California, the dominant tree is Douglas fir. During its lifetime of a thousand years or more, this magnificent tree may exceed ten feet in diameter and occasionally pushes to 300 feet in height. The deeply furrowed, reddish-brown bark of old-growth specimens make this monarch instantly recognizable. Its cones are equally distinctive, bearing a sort of forked tongue like that of the serpent that tempted Eve. As a sapling, Douglas fir makes a perfectly proportioned Christmas tree. The crown of the mature tree, whether growing in the forest or in the full sunlight, takes on an irregular and highly individual shape. A single stem or "leader" points straight up from the top of the tree; the fingerlike tip of each major branch aims outward and upward.

Ironically, Douglas fir isn't a fir at all. Its Latin name, *Pseudotsuga menziesii*, identifies it as a false hemlock. Early botanists (including David Douglas himself) mislabeled it variously as pine, hemlock, fir, and spruce. Only in the late nineteenth century, after discovering a related Asian species, did scientists conclude the tree represented a new genus, and the name *Pseudotsuga* stuck.

Nowhere in what foresters call the "Douglas fir region" is the tree a climax species. As a forest grows and matures, its vegetation changes until it reaches its climax, or final stage. Old growth, like the Douglas fir that so often dominates it, is a transitional stage. Intolerant of shade, Douglas fir can't grow in its own shadow; it must give way to more tolerant species. Fir manages to maintain its overall dominance because of its longevity, because thick bark sees it through most fires unscathed, and because it aggressively establishes itself in openings created by fire or other natural events.

Immense forests of giant Douglas fir standing butt to butt are no longer abundant as they were during the days when John Muir visited Puget Sound. Still, his observations on the tree he called Douglas spruce tell us much about a land that could give rise to a profusion of these giants:

> For so large a tree it is astonishing how many find nourishment and space to grow on any given area. The magnificent shafts push their spires into the sky close together with as regular a growth as that of a well-tilled field of grain. And no ground has been better tilled for the growth of trees than that on which these forests are growing. For it has been thoroughly ploughed and rolled by the mighty glaciers from the mountains, and sifted and mellowed and outspread in beds hundreds of feet in depth by the broad streams that issued from their fronts at the time of their recession, after they had long covered the land.

Like Douglas fir, western red cedar (*Thuja plicata*) is something other than what its name implies. An arborvitae rather than a true cedar, this majestic tree belongs to the cypress family. The tree's decay-resistant heartwood is as prized by shake splitters and fence builders today as it was by the Coast Salish natives a millennium ago. Unlike Douglas fir, shade-tolerant red cedar can maintain its forest dominance over the centuries. As long-lived and as massive at the base as fir--but not as tall--western red cedar is distinguished by its scaly leaves, its shaggy fir, and its flared, amoeba-shaped butt. Alaska yellow cedar, a smaller cousin of red cedar, may live up to 3,500 years.

Western hemlock (*Tsuga heterophylla*) had little commercial value until the pulp industry moved into the Northwest. Now used for lumber as well as paper and cellophane products, old-growth hemlock is especially popular in Japan. The pliable tip of hemlock, whether young or old, bends earthward as if in prayer. It's a lovely understory tree; its short needles spread flatly from the stem, forming a fanlike network of lace. This lacework of hemlock is a classic feature of a fir-hemlock forest.

Because this tree doesn't attain the mighty dimensions of Douglas fir or cedar, it's easy to dismiss it. But hemlock is a survivor. If a forest makes it through enough centuries without fire or other environmental catastrophe, this shade-tolerant tree will replace Douglas fir as the dominant tree. Hemlock is the most widespread climax species, ranging from the spruce forests of Alaska to the redwood country of California.

Sitka spruce (*Picea sitchensis*) thrives primarily in the coastal fog belt from southern Alaska to the southern tip of Oregon. Sitka spruce is an extremely fast-growing tree under the proper conditions. One thirteen-foot-thick Olympic rain forest specimen reportedly added a foot to its diameter in less than thirty-five years. Like hemlock, spruce prefers to begin life in the nurturing climate of a nurse log. Once used to build aircraft and now valued as a superior wood for piano sounding boards, Sitka spruce is Alaska's most important timber species.

The mighty coast redwood (*Sequoia sempervirens*) hugs the immediate coastal region from the edge of spruce country in southernmost Oregon south to the San Francisco Bay area. The redwood is the world's tallest tree, exceeded in mass only by the incomprehensibly large giant sequoia (*Sequoiadendron giganteum*) of the Sierra Nevada. Redwood appears to be a climax species--although some scientists question whether it, like cedar, eventually gives way to hemlock through the process of forest succession. Like red cedar, redwood is remarkably resistant to rot. Virgin redwood stands have been reduced to a fraction of their original extent, and heavy logging continues.

Big trees are the key ingredient in building the forest structure that has come to be called old growth. Yet big trees do not, by themselves, add up to an ecosystem. In fact, the trees of old growth are a varied lot. Just about the only thing this ecosystem lacks is uniformity. Its living trees differ by species, by age, by size, by shape and depth of their foliage, and by soundness of their bark and wood. Centuries are required for development of the distinctive multilayered canopy that comes with a mix of young and old trees. With time, older trees die, leaving their large remains for bacteria, fungi, insects, birds, and mammals to forage in and build homes.

Seen from a steep hillside above the forest floor or from a low-flying airplane, the irregular canopy of an ancient forest is striking. In contrast to the uniform, unbroken canopy of a second-growth forest, the old-growth canopy appears random and disordered. The dominant trees vary in height, thickness, and foliage. The tops of some trees are broken, others have lost their foliage and died, still others have fallen. The gaps left by these fallen or humbled comrades allow light to filter through to the understory. The crown of an old-growth tree has been described as resembling a bottle brush--"albeit one with many missing bristles."

Early settlers and loggers, working on level ground, had little opportunity to view the remarkable forest canopy from treetop level. But today, with

Above: Snags and live trees of varied species, age and size produce the rich habitat of the old-growth forest canopy. This scene is on the North Branch of the North Fork Stillaguamish River near Darrington, Washington. Right: The northern spotted owl, held here by researchers Eric Forsman (left) and Stan Sovern, has become a symbol of the ancient forest ecosystem.

most of the remaining old growth limited to narrow mountain valleys, the steep mountainsides offer wondrous panoramas. At some point almost every trail through ancient forest ascends to a high point from which it's possible to look through the canopy. The tops of the trees--many of them broken off, some alive, others dead, a few charred by lightning--may seem close enough to touch. To view the canopy from this perspective is as to feel as though one is actually in the canopy. The varying heights of the trees, their irregular spacing, the crowns of every shape and density, the quality of light, all create a more-real-than-real, super-three-dimensional effect.

Draped over the trees of old growth are mosses, lichens, club mosses, and liverworts. These "epiphytes," from the Greek prefix *epi* (upon) and suffix *phyte* (plant), grow nonparasitically atop other plants. Drawing their nutrients directly from rainfall and from particles in the air, epiphytes contribute significantly to the fertility of the forest. More than 130 epiphytic mosses have been identified in the Olympic rain forest alone. Lichens--symbiotic combinations of fungi and algae--may display tiny, delicate-looking branches or they may have a crude crustlike appearance. The most abundant epiphytic lichen is the large, leafy *Lobaria oregana*. As pieces of this and the closely related lungwort (*Lobaria pulmonaria*) break off and fall to the forest floor, they are eaten by browsing animals or they decompose to add valuable nitrogen to the soil. The waxy lungwort is named for its resemblance to the inside of the human lung. These nitrogen-fixing lichens are rare in young forests.

Until recently, it was believed that mosses and lichens had no directly beneficial or harmful effect on their host trees. It turns out, however, that some trees are more than passive hosts for the epiphytes. Nalini M. Nadkarni, a University of Washington graduate student in forest ecology, made a remarkable discovery when she used mountain climbing equipment to propel herself into the canopy of the Olympic rain forest. Weighing the epiphytes, she found three tons of dry plant mass per acre, four times the amount typical of other forests. Peeling back the vegetative mats, Nadkarni found a network of roots running along the branches and trunk of bigleaf maple. These maple roots, ranging from tiny white tips to chunky three-inch-diameter roots, were present only under moss and lichens. "Greatly excited," she wrote in *Natural History*, "I realized that here was a shortcut in the rain forest's nutrient cycling system. Host trees were capable of tapping the arboreal cupboards in their own crowns."

Nadkarni found that other deciduous trees in the Olympic rain forest--vine maple, red alder, and black cottonwood--also took water and nutrients from the organic matter created by epiphytes. Traveling to the cloud forests of Costa Rica and Papua New Guinea, she saw the same phenomenon of trees taking advantage of the nutrient treasure provided by mosses, liverworts, and lichens.

The old-growth canopy, developed to make the most of the sun's energy, has been likened to the ocean's surface. Zoologist Marston Bates put it this way in his classic book, *The Forest and the Sea*: "Most life is near the top, because that is where the sunlight strikes and everything below depends on this surface. Life in both the forest and the sea is distributed in horizontal layers." The crowns of trees are like the pelagic, or surface, layer of the ocean because that is where photosynthesis is concentrated. On the forest floor, as on the benthos, or bottom, of the sea, life depends on "second-hand materials that drift down from above--on fallen leaves, on fallen fruits, on roots and logs. Only a few special kinds of green plants were able to grow in the rather dim light that reached the forest floor."

Because scientists have not had 200 years to watch old-growth forests develop, they must deduce how the process works. The 1984 report of a Society of American Foresters (SAF) task force on old growth recognized these uncertainties. "Can old growth be managed?" the task force asked. The group answered its own question this way:

> Through silviculture, foresters can grow big trees and grow them faster than nature unassisted. Yet there is no evidence that old-growth conditions can be reproduced silviculturally. In fact, the question is essentially moot, as it would take 200 years or more to find an answer. Old-growth management, for the foreseeable future, will be predicated on preservation of existing old-growth stands.

Preservation may not be the most exact word for protecting a living system from destruction. If one fact emerges clearly from the new research on the forest ecosystem, it is that change is its very essence. The forest is forever in the process of *becoming*. Conifers are constantly struggling through the brush layer of a new clearing. Live trees keep being turned into fodder for the fungi that help to grow new trees. Like Sisyphus forever trying to reach the top of the hill, the forest continually struggles toward climax, the final and most stable stage of vegetative succession. Some natural or human-caused event inevitably sets the forest back to an earlier stage of succession, so old growth is constantly being created.

An old-growth forest "preserved" by humans inevitably will change. Old trees will fall and young trees will grow. Openings will be created and then filled with new life. The forest may reach climax. Or it may be completely toppled by windstorm or fire. Over a century, a 600-year-old stand may become a 700-year-old stand or a 100-year-old stand. But along with change, the forest maintains continuity and achieves a kind of stability. The young science of ecology teaches that complexity creates stability. The old-growth forests of the Northwest offer a wonderful laboratory--certainly the best on this continent--for studying the hidden ways in which seemingly independent creatures of the woods support one another.

The SAF task force wasn't saying that *all* old growth should be preserved.

The point was that in our hubris we shouldn't assume that we can duplicate a natural process we barely understand. Not surprisingly, one of the eleven members of the task force was Jerry Franklin.

The implications of the new forest research reach far beyond the arcane interests of the scientists. The productivity of tree farms, for example, may depend more on maintaining healthy wildlife populations than on eliminating "pests." Consider the class of fungi known as mycorrhizae. The growth of conifers, like that of many other photosynthetic plants, requires the assistance of these fungi whose name means "fungus root." Unable to make direct use of the sun's energy, a fungus attaches itself to tree roots and helps itself to the sugars produced by the host plant. In return, the fungus absorbs water and nutrients, passing them along to the host tree. Antibiotics in the fungus protect the tree from root-rot pathogens, while the mycorrhizal sheath around the roots creates a physical barrier against some parasites. Tons of mycorrhizae are present in a single acre of old-growth forest, feeding the trees and storing nutrients that otherwise would be washed out of the soil and lost to the forest. The contribution that mycorrhizal fungi make to tree growth and forest fertility can scarcely be overstressed.

The story of the mycorrhiza-tree relationship doesn't end there. What at first seems a *pas de deux* is in fact a *ménage à trois*. For without the small mammals of the forest, the symbiosis would break down. Hypogeous, or underground, fungi depend on mammals for much of the dispersal of their spores. Truffles, the underground fruiting bodies of these fungi, are an important food source for a number of small animals in the old-growth forest. An animal finds a truffle by smell, digs it up, and eats it. All of the truffle is digested--except the spores, which are excreted in the animal's feces. Wherever the animal goes, the fungus goes.

Truffles were the key to a mystery that had long baffled biologists. Biologists knew that bobcats and coyotes preyed on the northern flying squirrel. How do they catch an animal that presumably spent its time in the forest canopy? Analysis of the squirrels' stomach contents provided the answer. During the winter, lichens in tree crowns were the primary food source for flying squirrels. But between spring and fall, when the the snow-pack melted, the nocturnal animals came down to the ground to dig truffles. That's when their predators caught them. Laboratory studies of truffle-eating squirrels show that they pass nitrogen-fixing bacteria through their bodies along with fungal spores. Hypogeous fungi provide a home and small mammals a transport system for bacteria that add fertility to the forest floor.

The small California red-backed vole, whose tunnels are ubiquitous in the soil around rotting logs, shows a similar preference for truffles. The vole's environment, the Coast Range of southern Oregon and northern California, ·is ideal; this is one of very few places in the world where hypogeous fungi fruit year-round. If the fungus supply briefly runs short, the vole temporarily switches to lichens. Its favorite food may be *Rhizopogon vinicolor*, a mycorrhizal fungus that fruits mostly in rotten wood. *Rhizopogon* attaches itself to the roots of trees that grow in nurse logs. Forest researchers Chris Maser and James M. Trappe sum up the vole-fungus-tree relationship:

> Thus, there is a tight cycle of interdependence: the vole needs the truffle for food; the truffle depends on the vole for dispersal of spores and on a mycorrhizal tree host for energy; the tree requires mycorrhizal fungi for uptake of nutrients and provides the rotten wood needed by the vole for cover. Moreover, since both voles and *Rhizopogon vinicolor* specialize in rotten wood as habitat, the vole disperses the *Rhizopogon* spores to the kind of substrate in which the fungus thrives.

Are the small truffle-eaters the forester's friend or foe? Townsend's chipmunk and the deer mouse, long cursed as pests that eat conifer seeds, also are eager consumers of truffles. In that second role, they assist the growth of young Douglas fir by bringing mycorrhizal fungi to their roots. Maser, raising one of the criticial questions for forestry in the twenty-first century, argues that losses of wildlife and changes in soil chemistry may lead to declines in commercial forest productivity. Already, he believes, there are signs that this has happened in parts of Europe. Other scientists such as Roy R. Silen worry that the genetic diversity of the trees may be narrowed perilously far by clearcutting the virgin forests and replanting "genetically improved" trees.

In the long run, if not the short run, what's good for nature as a whole is good for the human species. The future of the timber industry depends on the hardiness and genetic diversity of the tree kingdom. Mature and old-growth forests are the reservoirs of that diversity. This era of impending climatic change would seem to be the least appropriate time to deliberately narrow the forests' genetic diversity.

More broadly viewed, the genetic variation within a particular species is only one part of the richness of life forms known as biological diversity. The other two elements of biological diversity are a variety of species and a variety of ecosystems. All three elements of genetic diversity are placed at risk as North America's last virgin forests are replaced by the even-aged monoculture of tree farms. What's happening to the forests of the Pacific Northwest mirrors a worldwide phenomenon. Global deforestation may cause the extinction of one-fourth of the earth's five million species over the next several decades. The use of plants from the tropical rain forests as sources of pharmaceuticals has been widely discussed. Only recently has the bark of the Pacific yew been identified as a potent weapon against cancerous tumors in mice. The National Cancer Institute in 1987 contracted for the harvest of 60,000 pounds of the yew's papery red bark to determine whether the taxol derivative is safe for use in humans. The yew, found in old-growth forests, is not grown on tree farms.

Jerry Franklin believes management practices on commercial forest-lands may be an even more important factor in maintaining biological diversity than is the amount of old growth that's preserved. No matter how successful environmentalists are in their efforts to save the ancient forests, he points out, commercially managed forests will cover a much greater area. If timber practices on those lands aren't changed, "we're going to effectively lose the war for biologic diversity." His recipe, and that of a growing number of his colleagues, is to keep more old-growth characteristics in the managed forests.

"We've got to quit leaving them in a billiard-table condition," says Franklin. "We have to leave more material behind, leave trees, snags, patches of reproduction, down logs--leaving a lot more heterogeneity in the cutover areas than we've been doing until this point. What we've got to do is leave our Germanic heritage behind us and cherish a little bit of disorder and chaos in our cutover areas."

During the short time that scientists have paid serious attention to the old-growth ecosystem, they have learned that the spotted owl is just one of dozens of vertebrates that are most at home in the ancient forests. They've learned about the tree vole, whose family may live in a single tree for generations; the northern flying squirrel, which feeds on lichens in the winter and truffles in the summer; the goshawk, whose short wings are adapted for flight among the trees; the pileated woodpecker, which excavates its home in large Douglas fir snags; the Olympic salamander, which lays its eggs in the rotting wood of large logs; the marbled murrelet, a seabird that returns each night to its nest in the ancient forest; and the national symbol, the bald eagle, which prefers to build its nest in older woods.

This growing recognition of the uniqueness of the old-growth ecosystem comes at an awkward time. It's awkward for the forest-products industry because the decline of private and state timber inventories leaves the industry more dependent than ever on federal sales of virgin timber. It's awkward for those concerned about the ecosystem because the amount of low-elevation old growth is dwindling fast. We've entered an era of scarcity, and public-policy decisions will be painful. For more than a century, the timber industry has been the leading employer in the Pacific Northwest. Although aerospace-industry employment has inched ahead in Washington and tourism is growing rapidly, the importance of timber can hardly be overstated. More than one-third of Oregon's manufacturing-sector jobs still are in lumber and wood products.

The challenge is to find ways of maintaining the economic vitality of the Northwest while preserving an irreplaceable national treasure. This book attempts to delineate the values of old-growth forests, to identify conflicts over their management, and to explore creative solutions. The focus is on the area of Oregon and Washington west of the Cascade Mountain crest, often called the Douglas fir region.

The more I have learned about the debate over the old growth, the more I have been struck by the complexity of the issues. We've gained some insight into the ancient forest ecosystem, but we can't quite put our finger on the glue that holds it together. We haven't taken seriously the need of Native Americans for pristine forests to practice their traditional religion. The timber industry contends that its tree farms could support a healthy population of spotted owls, but it still practices the kind of even-aged silviculture that drives owls out.

While independent sawmills are going out of business for lack of logs, two-fifths of Washington's timber harvest is sold abroad unmilled. The forest-products industry clamors for unabated timber sales but, for the most part, remains silent about the understocked second-growth forests and the conversion of the best timberlands to suburban tracts. The continuing loss of jobs in the forest-products industry has far more to do with the automation of sawmills, log exports, and conversion of private timberland than with measures designed to protect the old-growth forests.

We haven't figured out whether the regional economy benefits more from logging an old-growth forest or from promoting the recreational potential of the forest. Nor has anyone quantified the damage done to the Northwest's important salmon-fishing industry by logging and road building. Our growing knowledge about nutrient cycling in old-growth forests has raised disturbing questions about the sustainability of tree farming as currently practiced. Increasingly, it looks as though the future of the timber industry requires that old-growth characteristics be incorporated into tree farms.

The slash burning that typically follows clearcutting throws carbon into the atmosphere, exacerbating the greenhouse effect. Logging of the ancient forests is more than a regional problem. Virgin forests are falling around the world, from Vancouver Island to Indonesia and from the Amazon to Africa. Like the rain forests of the tropics, the temperate forests of Oregon and Washington are being stripped to keep a troubled industry afloat and to satisfy the voracious appetite of Japanese and American consumers for lovely, fine-grained wood.

The fate of the ancient forests has become the premier issue of public land management in the western United States. The basic question confronting the body politic is how much old growth should be cut for the timber industry's benefit and how much preserved. The courts have been asked to resolve the issue, but they appear to be only a detour on the way to Congress. This process may not yield the best results--and certainly won't if the issues remain as poorly understood as they are at present. A recent scene before a congressional subcommittee shouldn't surprise us. An environmentalist had been talking for some time about the importance of old-growth forests when a puzzled congressman interrupted to ask, "Okra? What's this okra you're talking about?"

What ought to concern us even more than the lack of knowledge of a decision-maker from the Southeast is the absence of constructive dialogue among those closest to the situation. Precious little problem-solving has taken place. One Seattle newspaper editor is reported to have thundered, "I hope I never read the word old growth again!" The Forest Service proposes, with none of Solomon's subtlety or wisdom, to carve the remaining old growth in two, giving half to the environmentalists and half to industry. As a solution it won't stick, because neither side is buying into it. When the courts and Congress get through deciding matters, the outcome may not do much to help an endangered ecosystem *or* industry. But an imposed settlement is exactly what's going to happen unless representatives of environmental groups and the timber industry decide there's more to be gained than lost by negotiation. A satisfactory settlement must be more creative than just carving lines through the old growth.

In desperate times, people turn to desperate measures. Environmental radicals have taken to sabotaging logging equipment and "inoculating" old-growth trees with spikes that pose a deadly hazard to mill workers. Not all loggers are kidding when they talk about hunting spotted owls.

Those who earn their livelihood from the forests feel they are under attack by outsiders who understand neither their way of life nor the forests themselves. It's as if Midwestern farmers, struggling to pay their debts, had to fight a powerful political movement bent on turning their farms back into a natural prairie. If environmentalists aren't a cynical elite, then, by golly, they must be naive sentimentalists duped by the Bambi syndrome.

The alleged sentimentality of environmentalists was confronted head-on by Robert Vincent, a wildlife consultant more concerned over the prospect of board feet of timber lost to industry than over the loss of spotted owl habitat. At a conference on old growth, Vincent noted that the richness and diversity of these forests is built on the death and decay of trees. He first likened old growth to a cemetery, then to "a self-centered miser" who buries his fortune in Crisco cans. On the other hand, he continued,

> ...there are some people who really enjoy old things. The unchanging, the stagnant, the dying. In fact, we almost seem to live in an era of ancestral worship of things. Where do people go to make a visit? Certainly not to Houston or Denver, but to Rome, Venice, Paris, and Brussels where they can see old cathedrals and stagnated cities that haven't been touched for years. In fact, the longer a city stagnates or the longer a ruin stands, the more beautiful it becomes. It's kind of ridiculous, isn't it? . . . I suppose that if we could preserve an automobile junkyard for 1,000 years, it might be considered a thing of beauty.

Ridiculous though it may seem to some, there is something in most of us that looks in awestruck wonder at a forest that has been growing stronger and wilder for a millennium. And, as those in the European tourism business know, people are willing to pay for that feeling of awe. There's money in them thar woods, whether they're chopped down or left standing. Our feelings of wonder at the ancient creations of man and nature go beyond dollars and cents, of course. As we stand at the last corner of our continent gazing at the fast-disappearing old growth, our values are being put to the test.

The determination of people in logging country to preserve their live-lihoods is a legitimate desire, one that must be confronted forthrightly. So, too, is the "sentimental" view that the virgin forests are a gift of God, not to be destroyed lightly.

Eric Forsman and Stan Sovern have more work to do. As soon as the spotted owl is banded, they clamber back to the truck and head out on the logging roads toward Forks. They're going to join another member of the research team, Timm Kaminski, who is trying to locate another bird and outfit her with a radio transmitter. If the effort is successful, the scientists will be able to track her movements through the winter and into the next year. The data they collect will add valuable information about the kinds of forest in which these owls nest and forage.

We follow a dirt road over state and private land that has been almost entirely cut over. Finally, we see the distinctive tree line of a natural forest. Most of these woods, Forsman explains, are "Twenty-one Blow." The hurricane that whipped through the west side of the Olympic Peninsula on January 21, 1921, tore down thousands, perhaps millions of old-growth trees. On one part of the peninsula, the 100-mile-an-hour winds left a thirty-foot-deep pile of trees; on another, an entire herd of 200 elk was wiped out by falling timber.

We pull off the road next to Kaminski's pickup. Forsman shouts to Kaminski, but there's no answer. So he shoulders a radio receiver and tunes into the channel on which the male owl's transmitter is broadcasting. By walking in on the male, he hopes to find Kaminski and the female he's after. The woods are remarkably open. Despite the young hemlocks growing beneath their mature elders, there are no impenetrable thickets here. Moss covers the ground, rotting logs support a profusion of new life, and the rotting base of a gigantic dead tree stands as a reminder of the forest that stood here before the Twenty-one Blow. This is not, strictly speaking, old growth. Like the first forest we visited, it's what Forsman calls a "mature" or "mixed stand." In common with old growth, this stand has a deep, multilayered canopy, a mottled pattern of soft sunlight, and an abundance of dead wood both standing and fallen. What it lacks are the very large trees that only centuries can produce. To see true old growth, we would have to cross the creek to a cedar grove with trees seven feet in diameter and larger. It's there, on a privately owned parcel surrounded by state land, that the birds nest.

Forsman picks up strong signals on his radio. Soon he spots Kaminski and sees the female roosting overhead. Sovern and I keep a respectful distance to avoid spooking the bird. Sovern goes through his pack and prepares his gear. The juvenile, which we can't see from our location, is begging its parents for food: *Whhhhhhhhhip!* . . . *Whhhhhhhhiip!* It sounds something like a preschooler trying to whistle. Forsman and Kaminski trap the female while Sovern continues to prepare the equipment needed to strap the radio to her back. The adult male, pumping his broad wings, flies above our heads toward his mate. The male and the juvenile watch from branches as Forsman makes the capture. The whole family is here.

Sovern and I walk over. Sovern takes hold of the bird. A feisty one, she snaps at Forsman as he painstakingly fits the tiny transmitter on her back and adjusts the thin straps that hold it on. With each snap of her powerful beak she makes a loud cracking sound, like the pop of two wood blocks being slammed together. Once, she manages to get Forsman's finger. The scientists cover her large brown eyes.

Not far away, the juvenile keeps begging for his dinner. The male protests his mate's treatment: *Hooo . . . hooo . . . hooo . . . hooo!*

The spotted owl is not, to some people's minds, the most intelligent of birds. Certainly it is not the most cautious. One spotted owl, oblivious to the human threat, wound up in Seattle's Woodland Park Zoo after being struck by a logging truck. The female in Sovern's hands has been moused so many times she has become quite tame. Cautious though she may be about crossing a clearcut, where she could become a great horned owl's prey, she fails to appreciate the peril that humans pose. While she lies helplessly, her family can only watch and voice their useless protest.

Perhaps it's best that they don't understand their precarious future. The birds on the branches are no safer than the bird on the ground. Their fate, like that of the forest itself, is in our hands.

2

People of the Cedar

In the beginning was the forest. The forest and the water. At the front door of the first people's villages was the water. Whether it was salt water at the mouth of a river or a river itself, the water was the path through which the salmon would make their way each year to present themselves for sacrifice to humans. At the back door was the forest.

Like the white men who one day would follow, the first people would clear a place on the edge of the woods for their villages. Occasionally, they also would set fires to maintain the clearings in which they grew their second most important food, the camas plant. They may have used fire periodically to hunt game or to clear out brush from the forest and make it easier to travel on foot. But the forest, with its tall, straight trees and its abundant game, always was there. The first people's impact on the forest was limited.

Some of the natives of what is today western Washington and western Oregon were true forest dwellers whose villages were surrounded by the deep woods. Most lived *in* the forest no more than they lived in the water. They lived on the forest's edge, their homes facing salt water. But just as the water's life-giving salmon shaped their way of life, so too did the forest and its wealth. Of all the plants and creatures of the wood, one exceeded all others in importance: cedar.

"Tree of life" is a term that has been applied aptly to western red cedar and to a lesser extent to Alaska yellow cedar. Cedar was as important to the natives of the coastal Northwest as was the buffalo to the people of the Great Plains. It provided clothing, shelter, transportation, tools, ropes, art objects, medicine, and ceremonial objects. Babies were cradled and diapered in cedar; the dead were buried in it. The Tlingit people, who lived north of red cedar's range in present-day Alaska, traded with their Haida neighbors to the south for cedar canoes. Inhabitants in the southern end of cedar's range, where trees didn't grow so large, also imported canoes.

Every part of a cedar tree was used: wood, bark, roots, and foliage. For some purposes, such as basket making, a modest amount of bark or root would do the job. Even the planks used to side a house could be carved out of the side of a standing tree. But to build a canoe or carve a totem pole, there was no getting around the need to fell a large tree. Cutting a tree the traditional way, without saws or metal axes, was a tedious process that took many days. Among the Makah of the Olympic Peninsula, ethnologist Hilary Stewart reports, only a wealthy family with many slaves would fell a large tree. The fallers would use wood-and-stone adzes, hacking at the trunk until the tree fell. In other parts of the Northwest, fire would be used, or a combination of chisel, wedge, and maul.

Shaping a canoe was an even more time-consuming process. Stewart, based on her interviews with traditional canoe-builders and observation of their work, has documented their methods. After the tree was cut, the canoe-maker would split a log to make two large canoes or three smaller ones. He would rough out the boat, and then allow the wood to cure over the winter. In the spring, the builder and his assistant would hollow out the log, either by chiseling it or by burning it out with hot rocks. With an adze, the master builder would carefully do the final shaping. The job still was not done. The sides of the canoe had to be spread by pouring water into the bottom and dropping in hot rocks to make steam. This process was not without risk. Ideal though cedar was for canoe-building, the wood would split sometimes under the strain of being hollowed out and expanded.

What made cedar such a sought-after wood? The attributes that make it so valuable as a commercial product today are precisely the qualities that gave it special meaning to the natives. The oils in western red cedar make it one of the most rot-resistant woods found anywhere. This is an especially important attribute in the rainy, misty Pacific Northwest. The wood remains popular today for roofing, siding, decking, and fencing. The tree of life has other practical values. For one, it is simply a wonderful wood to work with. Old-growth cedar has a grain that is straight, tight, and knot-free. The wood is soft, to boot--ideal for splitting into planks or shakes. As the Indians discovered thousands of years ago, cedar can be worked easily with the most basic of tools. Even today, shake mills split the wood in much the old way. The shaggy, grainy bark of the cedar is unique. It can be made into rope, skirts, mats, baskets and cooking pots, canoe bailers, and sails. The roots of the tree are used in basketry almost as much as the bark. For tribes on the coast of what is now California, redwood played much the same role as did cedar for their northern neighbors.

Native culture in the Northwest was not always based on cedar. As the environment changed, so did human culture. Recent scientific research has shown that the coastal forests have undergone major changes in the 12,000 years since the glaciers retreated north of Puget Sound. Layers of pollen laid down in peat bogs have given scientists a record of the changing vegetation west of the Cascade Mountains. As the glaciers began to recede from the southern tip of the sound, pine forests sprang up almost in their shadows. For 3,000 to 4,000 years, lodgepole pine dominated the Puget Trough and the Willamette Valley. Its dominance couldn't last long. Adapted to very dry conditions, lodgepole pine gave way to the much longer-lived Douglas fir. Even as Douglas fir became the preeminent tree of the Pacific Northwest, its decline began. More shade-tolerant species, most notably hemlock and cedar, slowly fought their way to dominance. As measured by pollen production, hemlock has been the predominant tree for several thousand years.

When the first people arrived in what was to become the Douglas fir region, they found a forested landscape in which lodgepole pine was giving way to Douglas fir as the dominant tree. Scientists do not know with certainty when humans arrived on the West Coast of the present-day United States. People have been in the Columbia River Basin for more than 8,000 years; they may have been there and on the coast considerably longer than that. After crossing the Bering land bridge, the first settlers may have squeezed through a narrow gap between two massive inland ice sheets. Or they may have taken the potentially more hospitable coastal route. A lower sea level at the time meant the coastal strip was wider than today. And because it apparently was not covered with glaciers, the strip may have been covered with old-growth forests much like those of today.

Over thousands of years of human habitation, old-growth fir and cedar became the forest giants. The development of a mature forest apparently played a key role in the maturation of human culture. In a provocative 1984 article in *Science*, Richard J. Hebda and Rolf W. Mathewes theorized that the coastal Indians' cultural development was closely correlated with the expansion of western red cedar. Between 10,000 and 6,000 years ago, the pollen record suggests, cedar was relatively rare. Perhaps partly in response to a moister climate, the abundance of cedar increased until the tree shared "codominant status" with western hemlock.

Hebda and Mathewes hypothesized that as big cedar trees became abundant around 5,000 years ago, the technology for working with this new resource began to develop. "Massive timber woodworking was probably well established by 3,000 to 3,500 years ago and reached a peak beginning 2,500 years ago and continuing to the present," the scientists reported.

> The maximums of the cedar pollen curves 2,000 to 5,000 years ago . . . and the development of massive timber working . . . appear to be closely correlated. We suggest that it was only during the latter part of the rise in the cedar curves that mature, large trees suitable for plank houses, canoes, and totem poles became available.

Ethnographer Wayne Suttles suggests that cultural and language differences between the Coast Salish tribes of the Puget Sound area and their neighbors to the north may be related to the predominance of Douglas fir in the south and of cedar, hemlock, and spruce in the north. (The linguistically and culturally related Coast Salish lived between Washington's Willapa Bay and British Columbia's Strait of Georgia.)

With its gifts of cedar and salmon, nature was kind to the natives of the Pacific Northwest. Although hunger probably was not unknown, it was a land of plenty. So plentiful, in fact, that recent archaeological excavations suggest the population of the Northwest coast between northern California and southeast Alaska numbered in the millions.

It was natural enough that a society so dependent on cedar would venerate the tree. But the affection felt by Northwest natives for cedar may be due to more than strictly functional reasons. Cedar has a melancholy, distinctively Northwestern sort of beauty. Even in a forest composed almost exclusively of conifers, cedar stands out. The austerely vertical lines of its bark--a sun-bleached grayish, white bark in older trees--conveys the feeling of a steady drizzle. The soft, bushlike foliage mirrors the bark with a rainlike droop of its own. The effect is one of sad beauty, an effect not lost on those whose very lives once depended on the tree. The Hesquiat Indians of Vancouver Island explained the loveliness of yellow cedar in a story of three young women who were drying salmon on the beach when the trickster Raven scared them into running up the mountainside. Breathless, the women stopped to rest and were turned into cedars.

Scholars debate one another over the extent to which the natives put their stamp on the forest environment. One author early in this century called the Indian "by nature an incendiary," and said forest-burning was "the Virginia Indian's besetting sin." Natives of the East and West coasts indeed set fires in and around the forests. They maintained openings for their villages and crop fields, and set fires to improve visibility and to make travel easier. These controlled burns favored regrowth of the kind of brush and young trees on which game prefer to browse.

Native Americans were neither irresponsible pyromaniacs nor noble savages who submitted passively to the whims of nature. Under their steward- ship, the forests of this continent probably were mixed--large tracts of un- touched, virgin forests interspersed with lightly managed forests close to settlements. The natives didn't hesitate to modify the landscape when they felt it was important to do so. Yet they lived lightly on the land so that it supported their lifestyle for many thousands of years. When white settlers arrived on Whidbey Island, the largest island in Puget Sound, they found prairies filled with bracken and camas. Coast Salish Indians had cultivated the plants, which were an important part of their diet.

"There can be successful alterations in natural systems," observed Richard White, in his excellent study of native and white land use on Whidbey and Camano islands. "The Salish changed the landscape and created an environment that remained stable and maintained the culture that created it over a long period of time without any weakening of its productive capability. Not all white changes proved so successful."

The original settlers of North America perceived their relationship to nature very differently than did the Europeans who followed. Whites settlers nearly killed the last of the buffalo for sport. The natives, by contrast, believed the salmon would return only if people performed the appropriate rituals.

Plants, animals, even the land itself, were perceived as the spiritual equals of humans. This world view, the foundation both of the natives' material culture and their religion, has not disappeared. It has brought the Coast Salish tribes of Puget Sound into sharp conflict with managers of the national forests.

Only a few people sit on the folding chairs circling the gym floor as a middle-aged woman rises from her seat. Dressed in casual clothes, she would look perfectly in place at a church social or at the local shopping mall--except for her face paint. On each cheek are three broad bands of black paint. She gazes into space and suddenly lets loose with a piercing cry that fills the large hall as though it were a mere shower stall.

"Uhhhh! Uhhhh! Uhhhh! Uhh-uhhhhh!" she chants over and over. Three cries, each the same pitch, are followed by a fourth cry that starts the same and drops to a deeper tone. The raw sound, bearing no resemblance to the artifice of Western singing, cuts through the centuries. The folded-up basketball net and the bright free-throw line beneath the glossy floor belie the transformation that is taking place. This is no longer twentieth-century America, a short drive from Interstate 5. This is another place, another time.

A group of men and women, their faces also covered with black or red paint, gather around the singer. Most wear sneakers, the men clad in blue jeans, the women in simple skirts. After the singer delivers several repetitions of her song, the others begin to bang on deerskin drums and they take up the song. Now the woman shifts from singing to dancing. To the accompaniment of her own song, her feet weave a seemingly weightless pattern on the floor, her hands gently fluttering by her side.

When she takes her seat, the spirit continues to move through her. She is wailing, her voice rising and falling like a siren. Near her, a younger woman, her face covered with black paint, begins to wail. The two voices fill the air. Slowly, the voice of the first woman dies away. The other stands to sing her song and do her dance.

It's an autumn afternoon on the Lummi Indian Reservation, where spirit dancers from around Puget Sound and Canada's Fraser River are gathering to remember one of their own. The family of Stella Long, a longtime spirit dancer, has invited members of the *seyowen* community, or "the smoke-house," to this memorial. Although two other *seyowen* gatherings are taking place at neighboring reservations this night, the family has chosen the tribal gymnasium because more people are expected than can fit into the cedar longhouse beside the Nooksack River.

Over the next few hours, perhaps 250 spirit dancers and their friends and relatives filter in. The warm-up dances continue. Two, even three, dances are going at a time. The cacophony is tremendous. A circle of drumming, chanting people surrounds each dancer. The chant and the beat produced by each group is different. Each dancer's guardian spirit has taught him or her a unique song. The dances are as individual as the songs. One, delicate and restrained, seems to trace the pulse of a bird's wings. Another, more athletic, resembles the leaping of a frog. Male black-paint dancers typically start in their seats, shaking their canes with deer-hoof rattlers, rocking to and fro and bellowing as they invoke their spirits. Some dancers, still in a trance, must be helped back to their seats.

Indians young and old wander in and out of the gym socializing and sharing a potluck feast in the dining hall. Guests have brought salmon, venison, and innumerable side dishes. The dancing continues for hours without pause. Late in the evening, attention turns to "the work." The women in Stella Long's family make a ceremonial entrance as they sing a family song to bring forth the woman's spirit. Her dancing regalia are brought out before the community for the last time, and her song is sung for the last time.

The work continues, with the family designating "witnesses" whose job it is to watch and remember each part of the ceremony. Helpers carry photographs of Stella Long and her late husband around the gym for all to see. "Floor openers" bring onto the main floor large plastic bags full of blankets that are to be distributed in the ancient potlatch tradition. Stella had wanted to show her appreciation to her friends in *seyowen*, a speaker for the family explains, so the family will now do what she can't do herself. Gifts are given with great ceremonial flourish. In addition to blankets, linens and china are handed out, along with money and handmade drums and quilts. Long speeches are delivered about the deceased couple and about the importance of *seyowen* and other Indian traditions. One speech is in the native tongue. After each ritual, a dozen family members line up to hand money to each witness.

Some of the teenagers in the bleachers grow restless during the lengthy "work." Others, seated on the edge of the gym floor with their parents and grandparents, eagerly await the resumption of dancing. Excitement fills the air as the first full dance of the night begins. The host family has asked four dancers representing different tribes and traditions to begin the dancing. The first dancer is a woman wearing a large headdress of cedar bark that suggests a bird. She sings her song, which is then taken up by a hundred drums and a hundred voices. Accompanied by her own song, she dances lightly, turning around, dipping and rising, making her way slowly around the large dance floor.

The effect is overpowering. What is so remarkable about this age-old ceremony isn't just the song or the circle of singing, drumming dancers--curiously moving though those are. Even more, it's the interplay of individual and group. With the help of the smokehouse community, this woman has brought forth the song of her spirit. As she embraces the song, the group embraces her. Individual and community are affirmed as part of a greater whole. To use an admittedly inadequate analogy, it's as though the members of a Catholic congregation were not only to take turns celebrating mass, but to write their own liturgies as well. Each dancer, in this public ritual, is celebrating a personal relationship with his or her own guardian power.

Tonight's powwow combines two previously separate traditions of the Coast Salish tribes: spirit dancing and the potlatch. The dancing that takes place in the smokehouse or, in this case, the gym, is the public, community aspect of *seyowen*. There's a private aspect as well, which has to do with the relationship between the dancer and the dancer's *seleya*, translated as one's power or guardian spirit. From this power, one receives a vision and a song, or *seyowen*. Traditionally, this vision was associated with a special skill such as hunting, fishing, or canoe building.

In the old days, a child was expected to go on a quest to receive a vision. During years of training, the child learned to travel through the forest in the dark, fast, bathe in winter waters, meditate, and, above all, to encounter one's power. In some training programs, a parent would hide an object deep in the woods and the child would be expected to find it. When the child was ready for the vision quest, he would make a solitary journey into the wilderness. The best areas were virgin spruce and cedar forests. The quester would sleep on spruce boughs and scrub with cedar. One's power could be associated with a plant or animal, even with an inanimate object (inanimate, at least, to the Western way of thinking) such as a lake, a waterfall, or a mountain.

The nature of one's vision and the identity of one's *seleya* is a private knowledge, not to be divulged. The story is told in northern Oregon of Plain Feather, a young man whose guardian spirit, an elk, made him a great hunter. For a long time, the hunter followed his spirit's advice not to kill more than he needed. But one fall, tricked by another person into expecting a hard winter, he killed to excess. He had broken his covenant with the spirit world. On one brutal hunt, Plain Feather followed a wounded elk to the shore of a small lake. The elk, which turned out to be his spirit, drew him into the lake. There, he saw the spirits of the animals he had killed, all in human form. Castigated by his spirit for his disobedience, he was thrown back out of the lake. Sick in body and spirit, he made his way back to the village where he told what had happened. Having revealed the identity of his *seleya*, he quickly died. From then on, the lake in which he saw the spirits was known as the Lake of the Lost Spirits.

Novice spirit dancers no longer make a solitary journey into the wilderness to find their power. Instead, they typically go through a ten-day initiation process inside the smokehouse. According to one account, the initiation has been altered because the wild places where spirits may be found have lost their purity and solitude. One black-paint dancer told me that passersby would become alarmed if they encountered initiates "wandering out there with no food or survival things." There might indeed be considerable risk these days to novice dancers who have had little experience in the wilderness, particularly in an age when the edge of the wilderness has receded so far from the reservations.

Yet the smokehouse religion maintains its link with the natural world.

The essence of *seyowen* still is the personal relationship with a guardian spirit rooted in the natural world. Virgin forests are still used for meditation and to gather ritual materials. Cedar is used in a shaman's regalia. Red-paint dancers make their hats of cedar bark. The sacred *sqwidilic* boards and *sqwidilic* headbands are made of cedar. Red paint is usually made from the flesh of a hard-to-find tree fungus, occasionally from ochre. Devil's club, wasp nests, or nettles are burned to make black face paint. Smokehouses are built of mature or old-growth cedar. At the end of the dance season, the "baby" poles of novice dancers must be hidden in the forest in a spot where they will not be disturbed.

Perhaps the most important--and most threatened--use of wilderness is the ritual bathing dancers do for spiritual purification. Like the dancing itself, this "swimming" is done during the winter dance season. If the dancer has to break through the ice, so much the better. For four years following initiation, dancers are expected to bathe every day. Before bathing, dancers fast. In the water, they shout to invoke the spirits and then pray to the four cardinal directions. Upon emerging, they rub themselves with cedar boughs and then run as fast as they can.

Finding appropriate places for ritual bathing has become a challenge. Key requirements are that the waist-deep stream or pond be pure and unpolluted, that there be total solitude, that the forest be unspoiled, and that there be cedar and ferns near the water. There was a time, before logging and human-caused fires leveled the low-lying forests, when dancers could walk from their longhouses to sites suitable for vision questing. Almost all the remaining questing sites are in the mountainous national forests, a one- to two-hour drive from the Lummi, Swinomish, and Tulalip reservations. Unless the site is near a road, the trip becomes prohibitively long. Elderly dancers, particularly, cannot be expected to undertake a strenuous hike into the rugged areas that have been designated as wilderness. At more accessible sites, spirit dancers are more likely to encounter backpackers or hunters. Questing sites may be moved as the water becomes too deep in one location or too shallow in another.

The conflict that disturbs spirit dancers the most is the clearcutting of the last virgin forests. One spot that was especially popular for ritual bathing, the Valley of the Eagles, was described by some Puget Sound Indians as a Yellowstone Park in miniature. A small stream tumbled down a series of waterfalls into a pool surrounded by cedar. "It was beautiful there," says Dave Jefferson, a spirit dancer and Lummi Tribal Council member. "A couple years ago they logged it all off. Now it's just a wasteland. It's all scrub brush." Spirit dancers had to move on to other questing sites higher in the mountains, farther from home. As logging roads are pushed into roadless areas, there are fewer places that can be used for spirit questing. In some of those spots, says another dancer, "you've got hippies on the side of the hills with binoculars."

The need for privacy has placed spirit dancers in a bind. The last thing they want to do is to encourage more gawkers by publicizing the locations of the spots where they meet their spirits. At the same time, they want the Forest Service to protect those sacred sites--something the agency says it can't do unless it knows where those sites are. In the Mount Baker-Snoqualmie National Forest, the uneasy solution to this dilemma has been for the tribes to identify broadly the areas that are most important for ritual purposes. Those areas are identified on maps provided to the agency on a confidential basis.

Before selling timber or building roads in those areas of concern, national forest officials have agreed to discuss their plans with Indian leaders and in some cases modify the plans. It's a process that apparently meets the requirements of the American Indian Religious Freedom Act of 1978 but leaves natives watching helplessly as the last of their sacred land is gradually whittled away. Ideally, spirit dancers would like their solitude protected through the designation of certain areas for their exclusive use. That isn't an approach the Forest Service is willing to take, saying the Indians can be guaranteed access but not exclusive use of sacred sites.

Some Puget Sound Indians explain their need for wilderness solitude in Christian terms. This isn't surprising, since some dancers belong to Catholic, Protestant, or Indian Shaker churches and see no conflict in their activities. Jesus, they recall, spent forty days in the wilderness before coming to terms with his divine mission. Lummi tribal councilman Clarence Bob likens old-growth trees to the Virgin Mary. "Would anybody come along and take her life before her time?" he asks. "That's how strongly I believe."

Sam Cagey, an elder in the smokehouse and a Lummi tribal councilman, grows angry when the subject of forest management comes up. "The white people are happy to have a few trees growing on an acre of land and they call it a park. To the Indians, the forest is their temple. We don't question the Mormon temple in Salt Lake City. We don't question the Vatican's temple. We don't want anybody to question how important nature is to us, because that's our temple."

It isn't difficult, even for an American of European descent, to appreciate the importance the Coast Salish place on natural forests as places of meditation. Visitors to the old growth often come away speaking of the "cathedral forest." It's an apt description. The architecture of these woods is so very Gothic that I sometimes wonder if the virgin forests of medieval Europe provided the blueprint for the early churches. Like the nave of a Gothic monument, a grove of massive cedars or firs is both sheltered and open. The air is still. Dappled light filters through the deep crowns of tall trees much like the subdued light from stained-glass windows. Sounds carry, but without the sharp echoes of a stone building. The vertical thrust of the large trees invites us to lift up our eyes. So accustomed are we these days to younger forests, with their thickets of slender trees, that it's striking to see how open an old forest is.

Above: Virgin cedar groves with trees like this 800-year-old specimen towering over spirit dancer Sam Cagey are central to the Northwest tribes' seyowen rituals. Right: Clayoquot Indians of Vancouver Island fashion a small dugout cedar canoe in the traditional manner.

For most whites, it may be no more than a metaphor to call the forest a cathedral. To Salish spirit dancers, following an age-old tradition, the forest truly is a place of worship.

Do the tribes of the West Coast have a First Amendment religious right to keep the last old-growth forests virgin? The answer to that question, as given by the U.S. Supreme Court, is no. The judicially decisive case began with the Forest Service's attempt to punch a logging road into a 31,500-acre roadless area in the Six Rivers National Forest of northern California. After more than a dozen years of planning, the Forest Service in 1981 proposed to open up that area of the Siskiyou Mountains to logging. The last six miles of a seventy-five-mile logging road were to be completed between Gasquet and Orleans, and 733 million board feet of virgin Douglas fir were to be cut over the next eighty years.

The legal battle against the G-O Road began in 1974 with an unsuccessful lawsuit by the Sierra Club. But over the following decade, opposition to the road became a *cause célèbre* not only for Sierra, Audubon, and Friends of the Earth, but also for three California Indian tribes. It was the involvement of the Yurok, Karok, and Talowa tribes that set the movement against the G-O Road apart from other environmental battles. The final road link would cut through their sacred "high country." Traditional Indians viewed the rock outcroppings of Chimney Rock, Doctor Rock, and Peak 8 as especially potent seats of power. Although only a limited number of natives made trips to those peaks, the pristine landscape was critical to those who did. The Indians' lawyers told the court: "The practices conducted in the high country entail intense meditation and require the practitioner to achieve a profound awareness of the natural environment. Prayer seats are oriented so there is an unobstructed view, and the practitioner must be surrounded by undisturbed naturalness."

The importance of the high country to the native religion was not unknown to the Forest Service. The Six Rivers National Forest had commissioned a study of Indian religious use of the forest. Completed in 1979, the Theodoratus Report pointed out the inseparability of religion from other aspects of life in native culture, and stressed the individual's relationship with the natural world as the underpinning of the culture. Construction of the G-O Road along any of the nine routes under consideration by the Forest Service "will produce an irreparable impact on the spiritual and physical well-being of the adjacent Yurok, Karok, and Tolowa communities," the report concluded. The cultural impact would be so damaging to the tribes, the consultants wrote, that the last link of the road shouldn't be built.

Too much was at stake for the Forest Service and its biggest timber customer in the area, Simpson Timber, to accept the Theodoratus recommendations. In its final 1982 plan for the Blue Creek area, the Forest Service announced its intention to go ahead with massive timber sales and road building in the high country. Half-mile-wide protective strips would be set up around identified religious sites. But there was no getting around the fact that meditation in the native prayer seats would be disturbed by the intrusion of bulldozers and logging trucks. The stage was set for all-out legal warfare.

Five years were required for the precedent-setting case to make its way from U.S. District Court to the highest court in the land. At the trial court level, the plaintiffs scored a resounding victory. The District Court ruled that the Forest Service's plans violated the Indians' First Amendment right to freedom of religion, the National Environmental Policy Act, the Wilderness Act, the Clean Water Act, Indian water and fishing rights, and the Adminis-trative Procedures Act. The court enjoined the Forest Service from building any roads or allowing any logging in the high country. The government was ordered to study the potential for wilderness designation of other parts of the Blue Creek Roadless Area and to consider additional measures to mitigate likely damage to streams and fisheries.

Of the Indians' religious claims, the trial court held that ritual use of the high country was "central and indispensable." Natives could communicate with the Great Creator only if the environment remained pristine, the court ruled. It was the first time a federal court had invoked the First Amendment to protect Indian sacred ground on federal land. In earlier lawsuits over the Glen Canyon and the Tellico dams, the courts had put the government's property rights ahead of Indians' religious claims. The legal precedent set in the high country case was so important that the San Francisco-based Ninth Circuit Court of Appeals heard the case not once but twice. Both times the appeals court upheld the lower court ruling. Construction of the G-O Road would "virtually destroy the Indians' ability to practice their religion," wrote the appeals panel.

The Indians' victory was short-lived. In its April 1988, the Supreme Court rendered its 5-3 decision in *Lyng v. Northwest Indian Cemetery Protective Association*, overturning the ruling of the district and appeals courts. The high court did not question the centrality of the pristine forest to the Indian religion:

> The Government does not dispute, and we have no reason to doubt, that the logging and road-building projects at issue in this case could have devastating effects on traditional Indian religious practices. The practices are intimately and inextricably bound up with the unique features of the Chimney Rock area, which is known to the Indians as the "high country."

But the opinion written by Justice Sandra Day O'Connor went on to note that Indian traditionalists weren't *prohibited* from visiting sacred locations; thus their First Amendment rights weren't threatened by logging or road-building. The key point, O'Connor wrote, was that freedom of religion should not be construed to give any religious group "a veto over public programs."

The government's property rights must take precedence. Justice William J. Brennan, in dissent, took the majority to task for its "astounding" opinion that the Constitution offers no practical protection of Native American religions.

The immediate effect of the decision on the Siskiyou high country was not clear. In 1984, while the litigation was pending, Congress had designated most of the high country as wilderness. That designation blocked logging. However, the wilderness act specifically excluded a narrow corridor that would allow construction of the G-O Road. Since the road would not open up any new areas to logging, it was possible the Forest Service would decide the road wasn't worth the cost. In the wake of the wilderness bill, the government's primary reason for pursuing the case was to establish its property rights over the Indians' religious-freedom rights.

The Rehnquist Court gave the government the legal green light it wanted. The message was clear: the Bill of Rights would not be interpreted in a way that might jeopardize billions of dollars' worth of timber and mineral sales, dams, and power plants. The religious needs of the first Americans were simply too inconvenient for society to accommodate. To be sure, the Supreme Court said its decision should not be taken as encouraging government insensitivity to native religions. But taken together with other legal setbacks, Indians were left virtually defenseless. The appeals court already had ruled that Indian traditionalists enjoyed no effective statutory protection. The American Indian Religious Freedom Act (AIRFA), passed in 1978, supposedly guaranteed "access" to traditional religious sites. But Congress had left the law deliberately vague. Rep. Morris Udall, sponsor of the House bill, pointedly told his colleagues that the bill "has no teeth in it." The Supreme Court agreed with that assessment, even quoting Udall to that effect. Indians were caught between a toothless AIRFA and an unsympathetic high court. If native religions were to be protected, it wouldn't be as a fundamental constitutional right. It would have to be through a new act of Congress or sensitive management by the executive branch of government. And neither was prepared to meet native culture on its own terms.

Four centuries after the European settlement of North America began, the gulf between the two cultures was as wide as ever. Contemporary Indians might wear blue jeans and drive Chevrolets, but their culture remained distinct. We shouldn't be surprised, observed historian Calvin Martin, that Native Americans continued losing their legal land claims. He cited a 1971 case in which the Crees of Quebec attempted to block the James Bay hydroelectric project that threatened to flood much of their hunting territory:

> It was a bizarre piece of litigation: members of a functioning hunting culture with elements going back thousands of years trying to explain, in a sterile Montreal courtroom, what that land signified to them. They were people from another world. . . .

Their petition rested its case on the following proposition: "We
... oppose ... these projects because we believe that only the
beavers had the right to build dams in our territory." The stunning
thing is that they were absolutely serious; this document was not
intended to be flippant or sentimental. For them, this was the
most cogent argument they could advance. And in their world it
was cogent indeed.

The Indian tribes of British Columbia have been more militant than the
tribes of Washington in their fight to save the forests. The Sitka spruce groves
of Vancouver Island and the cedar groves of the Queen Charlotte Islands are
so spectacular they make some of the Douglas fir region forests look almost
like second growth. On Lyell Island in the Queen Charlottes, Mounties
arrested more than 70 Haidas, including an 81-year-old woman, who block-
aded a logging road. The Indians' protest capped a ten-year struggle by environmentalists to save a chunk of the rain forest. The drama of Indian elders
submitting to arrest to save their homeland created a moral crisis. The result
was creation of South Moresby National Park.

That 1987 victory was overshadowed by the continuing pillage of the
Canadian forests. With little of the environmental protection or public review
that Americans take for granted, Canada's forests were being clearcut much
more rapidly than they could be replanted. Indians and environmentalists
focused their efforts on a few outstanding patches of virgin forests: Carmanah
Creek and Clayoquot Sound on Vancouver Island, and the inland Stein Valley.

With the possible exception of the California G-O Road, the most serious
threat to native religions from federal logging sales in the United States
probably was that posed by the Mount Baker-Snoqualmie National Forest's
timber sales. Because of the Salish Indians' extensive use of the Puget Sound
old-growth forests, tribal representatives pressured the Forest Service to
conduct an inventory identifying the religious needs. Only after such a study,
the Indians argued, would it be possible for the Mount Baker-Snoqualmie
Forest to ensure that they would be given access to religious sites as required
by the American Indian Religious Freedom Act. The Forest Service reluc-
tantly agreed to sponsor the study.

The results of the study coordinated by ethnologist Astrida R. Blukis
Onat were similar to the findings of the Theodoratus Report in the Six Rivers
National Forest. Blukis Onat's 1981 report documented religious use of the
forests by fifteen Puget Sound tribes. The report focused on the preservation
of "spirit sites" where questing and bathing are done, of plants used for
medicinal or ritual purposes, and of old-growth cedar stands.

The availability of cedar, Blukis Onat observed, has been "directly,
profoundly, and negatively affected by past management practices." The
consultant outlined the many traditional uses of cedar and noted, "In fact the
significance of cedar goes beyond that attributed to it by practitioners of the

traditional religion. It is considered by virtually all Indian people in the region as intimately linked to Indianness." Blukis Onat recommended that old-growth cedar stands be left intact throughout the forest except for tree-falling by the traditional religious community for the construction of smokehouses and other ritual purposes. It was a sweeping recommendation. By the time all the spirit sites, cedar groves, and other areas with ceremonial plants had been identified, some 450,000 acres were at issue. Some portion of that acreage already was protected in wilderness. But the amount recommended for protection in the Blukis Onat Report totaled more than one-third of the 1.3 million acres of forested land in the Mount Baker-Snoqualmie Forest. This tended to be the most valuable timberland in the forest. And it was concentrated in the northern half of the forest, which accounted for a disproportionately high percentage of timber sales.

Blukis Onat's inventory was based on Forest Service maps showing cedar stands and on Indians' reports of the sites they use. The mapping inventory was not precise. When it came to identifying spirit sites, spirit dancers were willing to identify only general areas rather than specific locations. Not only were some Indians concerned about possible breaches of confidentiality, there was the additional complication that spirit dancers' activities shift from one spot to another. Explains Juanita Jefferson, a dancer who acted as Lummi correspondent for the Blukis Onat study, "You can't put your finger on the map and say, 'That's a shrine or a sacred area, protect it.' The whole forest may be important to use one time or another."

That sort of argument doesn't go far, however, with a bureaucracy managing forest lands primarily for commodity sales. If Mount Baker-Snoqualmie National Forest officials had accepted the recommendations of Blukis Onat and the Salish tribes, the bulk of the forest's timber operations would have come to a screeching halt. The Forest Service took a politically safe approach that apparently met the requirements of AIRFA, if not the religious needs of Indians. The Forest Service would "consult" the affected tribes before selling timber or punching roads in areas identified as having religious significance. Timber sales might be relocated or put on hold to protect sensitive sites. But the agency would not agree to the blanket protection of spirit sites and cedar groves sought by the tribes and the Forest Service's consultant.

When the Mount Baker-Snoqualmie Forest released the draft of its long-range management plan in the winter of 1988, the level of protection offered to Indian religious uses was not even close to what the tribes felt they needed. The alternative proposed by the Forest Service would protect all known Indian cemeteries (300 acres of the 1.7-million-acre forest) and would protect two-thirds of the sites where native traditionalists gather ceremonial plants other than cedar. Where the plan was least supportive of the Salish religion was in its management of cedar groves and questing sites. The tribes had identified some 132,000 acres in which spirit dancers currently quest for spirits or have done so in the past; the Mount Baker-Snoqualmie Plan would give a "high" or "moderate" level of protection to just over half of those areas. Similarly, the plan would protect less than one-third of the 186,000 acres of cedar groves that weren't currently used for questing but which the Indians held essential to their religion.

To Doug MacWilliams, supervisor of the Mount Baker-Snoqualmie National Forest, the management plan represented one more step toward meeting Indians' rights under AIRFA as well as the Forest Service's "multiple-use" mandate. The name of the game was compromise--a compromise that sought to balance the Indian traditionalists' use of virgin forests with the timber industry's needs. MacWilliams insisted the consultation process had been successful, contending that agreement was reached on most timber sales when government and tribal representatives walked the land together to explore site-specific solutions.

A twenty-eight-year Forest Service officer whose primary training and experience was in commercial timber management, MacWilliams had been known to express skepticism about the legitimacy of the traditional Indian religion. In a 1985 article in *The Seattle Times*, he was quoted as saying *seyowen* has "been so watered down and diffused. . . . The chain has been broken. They have lost it." When I interviewed MacWilliams early in 1988, he acknowledged that the traditional religion is "real" to its followers and said he supported their practice of it. But, he added, the Indian community is no more monolithic than white society. Some Indian loggers, for instance, are more interested in cutting trees than in protecting spirit sites.

To the followers of *seyowen*, the Mount Baker-Snoqualmie Plan was alarming. All of the old-growth forest close to the native settlements around Puget Sound, or the *Whulj*, were long gone. Most of the remaining virgin forest, thirty miles or more from today's reservations, was crisscrossed by roads and hiking trails, much of it already fragmented by clearcutting. Much of the remainder was on the chopping block. Indians have watched helplessly while sacred sites such as the Valley of the Eagles were stripped of trees. They fully expect the same to happen to other spirit questing sites. Just west of Mount Baker, or *Koma Kulshan*, lies Warm Creek. In the consultation process, Lummi Indians told the Forest Service of the importance to them of this roadless area, but they saw the area earmarked for timber sales under the draft plan.

Kurt Russo, the impassioned forester who works for the Lummi Tribe and who organized the Puget Sound tribes to persuade the Forest Service to commission the Blukis Onat study, calls the consultation process "a sham." He describes the process:

> The Forest Service people will come to Lummi and say, 'Look, we're going to have to get this wood out. What we'll do is we'll distribute it over fifty years instead of twenty-five years. There's

our first compromise. What will you compromise?' And of course
the people are going, 'That's nothing.' My impression is it's a
compromise for the smokehouse people and the tribes to be in
those meetings in the first place. If you look at the history of how
the Forest Service got the land, they ought to be coming to the
tribes and asking, 'How should we be managing these?' not, 'Here's
what we're going to do and what do you think?' Being in these
consultation meetings is like being in *Alice in Wonderland*.
They'll come up and say, 'Let's compromise and consult.' Their
compromise is not whether but when.

To many members of the Salish tribes, the Forest Service's reluctance to
protect forests for religious purposes is but one more chapter in a long
history of cultural repression. Lummi elders still remember being taken from
their parents at an early age and put in boarding schools where they were
for-bidden from speaking their language or their parents' religion. Joe Washington, Sr., tells the depressingly familiar story of how he was whisked off to
a Catholic school at Fort Lapwai, Idaho, at age four. His teacher would burn
his tongue with a match when he spoke in the Straits Salish tongue, the only
language he knew.

While children were forced to attend Christian schools, the authorities
did what they could back on the reservations to stamp out the native culture.
In a typical report on Puget Sound's Indians, one government agent wrote:

From close observation I am satisfied that the greatest obstacle to
progress and to the advancement of the young Indian is the old
Indian. He still clings to his old superstitions and cherishes
secretly the old traditions and teachings of his savage ancestors.
He is opposed to sending his children to school; creates all the
dissatisfaction and distrust that he can secretly foment in the
child's mind; interferes with the agency physician in the treatment of patients; and does whatever he can in the two months of
vacation to neutralize the good effect of the ten months' school
session. With his disappearance from the scene of action, a more
rapid advance will take place among the younger Indians.

In 1921, the Office of Indian Affairs went so far as to make spirit dancing
and potlatches legally punishable "Indian offences." Joe Washington remembers standing watch outside his home while spirit dancing took place inside.
For some years, *seyowen* was abandoned on the Lummi Reservation. The
tradition was slowly revived and since the 1960s has grown dramatically.
Young people see it as a way of recovering their identity and many parents
send their children into the smokehouse as an antidote to such problems as
alcohol and drug abuse.

The Lummi Tribe is also attempting to revive the native language. Bill

James is putting together a dictionary of the language and teaching language classes to children and adults who have known no tongue but English. The ambitious goal of this tribal education project is to make Straits Salish once again the first language of a bilingual people. That's a tall order. James, a mountain of a man with a gentle voice like so many Indians, says there are "maybe five" Lummi elders alive who speak the language fluently. One is reminded of Israel's difficult but successful campaign to make the "dead" Hebrew language the language of everyday use. It was easier for Israel. The Jews returning to Israel spoke a mixture of languages, making it necessary for most of them to learn a new tongue anyway. They also were not fighting against the tide of a dominant culture whose language was the coin of the realm.

But like the Israelis of the 1940s and 1950s, the Salish Indians of the 1980s are working hard to rebuild a culture. Bill James and his mother, Fran, also teach traditional weaving and dyeing techniques. They discourage their students from using materials like raffia, which makes excellent baskets but is not native to the area. When gathering cedar bark, Bill teaches his students the importance of their relationship to the tree: "You talk to the trees and they talk to you. You ask the trees for their bark before you take it. You tell them what you want it for, that you're not going to kill them, that you just want them to share some bark with you."

The rebirth of *seyowen*, which now can count upwards of 5,000 adherents, is the most remarkable success in the cultural reconstruction. Native traditionalists are picking up growing political backing in their fight to save the virgin cedar groves. One is from the same Christian churches that tried to stamp out the native religion only a generation ago. More concerned now about freedom of religion in a pluralistic society, mainstream churches intervened in court on behalf of the California tribes fighting to save the Chimney Rock high country. And in Puget Sound, regional representatives of eight Protestant denominations and the Roman Catholic Church in 1987 issued a formal apology "on behalf of our churches for their long-standing participation in the destruction of traditional Native American spiritual practices." The bishops called on their denominations to help Indians gain protection of sacred sites on public lands.

Similarly, the environmental movement began to make common cause with the tribes. Spirit dancers and white conservationists perceive a common interest in protecting the last old-growth forests. Bill Arthur, regional representative of the Sierra Club, recalls a meeting in which tribal leaders were talking with Mount Baker-Snoqualmie National Forest officers over planning issues. When Arthur and another environmental activist arrived at the meeting late, Forest Supervisor Doug MacWilliams looked up and, according to Arthur, "It was like this gray pallor went over his face as he realized, 'Oh, God, the Indians and the environmentalists want to hug the same trees!'" Many conservationists, like traditional Indians, see nature as offering spiritual renewal to humanity. Similarly, both groups view human progress as something other than subduing nature.

Two remarkable facts stand out when one considers the Coast Salish Indians' culture and their special relationship to the forest. One is that the traditions have survived at all. The other is that the assault on the culture continues. Spirit dancers teach their children that their identity is linked with powers found in the wild places. From the Indian perspective those wild places, the forest temples of *seyowen*, are being looted.

3

Subduing the Wilderness

I t wasn't only the natives of this green continent for whom the forests were a central fact of life. The first European settlers were no less impressed by the woods that stretched almost unbroken for a thousand miles to the west. By the time they discovered the mightiest forests of all--the coastal rain forests of the West Coast--their appetite for timber was so voracious that the most accessible of these woods fell fast. No other nation of world-class size was built as quickly as the United States. Nor was any other nation's culture and economy based so completely on a single material: wood. Nowhere else was a forest of such dimensions subdued so quickly.

The types of woods that the settlers found were as diverse as the ethnic backgrounds of the nation-builders themselves. In the East, they found deciduous forests teeming with oak, cherry, hickory, and maple. The South was rich with fast-growing pines and subtropical swamps of cypress. In the northern woods from Maine to Wisconsin, the pioneers mined a wealth of white pine. On the West Coast grew the tallest, thickest trees: the giant sequoias, redwoods, and Douglas firs. Almost as quickly as the woods were discovered, they were cut down or burned off until the original stands of the East, the North, and the South were all gone. This land had given birth to a race of woodsmen.

Seventeenth- and eighteenth-century visitors to the New World were impressed by more than the untamed wilderness that surrounded the growing band of coastal settlements. They were struck, some favorably and some unfavorably, by the degree to which a whole society was being built with wood. Nearly every structure was built of wood, tools and implements were wooden, wood was burned in place of coal. Timber and forest products were becoming the leading exports. One English visitor grumbled about encountering "too many wooden houses for the Credit of the Place." Another spoke of "a Wooden Town in a Wooden Country & a wooden bred set of Tavern-keepers."

To build this "Wooden Country," the massive forest had to be leveled tree by tree. This logging upset some visitors. One Briton who traveled to the new nation opined that Americans "have an unconquerable aversion to trees . . . not one is spared; all share the same fate, and all are involved in the same general havoc." Many Americans and at least a few Europeans responded to such moral outrage by pointing out that a European-style society could not be carved out of the wilderness without first opening up the forest. Trees must be felled before crops could be planted and before cities could be raised on this forested continent.

A marvelous, if forbidding, forest it was. As Europeans explored the eastern coast of the Americas, they saw a sight that had disappeared from the Old World centuries earlier. Wherever their sailing ships set anchor or passed by, the forest was there. A continent, it seemed, just waiting for their settlement. They hardly knew what to make of such lush growth. Where Massachusetts Pilgrim William Bradford saw a wilderness "hideous and desolate," Italian explorer Giovanni da Verrazano gushed over forests "with as much beauty and delectable appearance as it would be possible to express." Christopher Columbus, apparently feeling less need to render judgment on the virtues of the forest, simply marveled at the tall trees "stretching to the stars with leaves never shed."

The first explorers and pioneers had no idea how vast the continent was. Nor could they know that its most spectacular forests grew on the far side of the land mass. But they wasted no time in setting to work. As quickly as men could swing their axes, the virgin forest began to fall--none too soon to suit the Pilgrims but perhaps with some regret for the likes of Verrazano. In time, the necessity of depending on wood became a love affair with it. Nowhere was this demonstrated more graphically than in San Diego's rapid transformation from an adobe town to a wooden city following its annexation by the United States. Instead of relying on local building materials in a forest-poor region, the settlers preferred to pay top dollar for lumber shipped from the virgin forests of the Pacific Northwest. When Simon Benson first began hauling log rafts 1,200 miles by ocean from the Columbia River to his San Diego sawmill, many scoffed at "Benson's folly." But after the city grew in the usual American wooden manner--to Benson's substantial benefit--the laughter died away.

In the late colonial period and early years of American independence, there was more than a small bit of jealousy at work in the disparaging remarks of some English visitors. After all, their country had severely depleted its royal forests in the late 1500s to build the fleet that defeated the Spanish Armada. In the decades that followed, the Crown sold off timber to produce a revenue independent of Parliament, and in the cruelest blow, thieves and creditors helped themselves to the royal trees. England turned more desperately than ever to the Baltic nations for timber supplies. But wars and the tough bargaining of Sweden's King Charles XII increased the British empire's desperation. Under those circumstances, it is no surprise that England chose to develop its North American colonies as a source of plentiful timber. England was not alone: Spain, France, and Holland were doing the same thing. The Jamestown settlement, in its first year, 1607, shipped a few handsawn clapboards to the mother country. Within four years a water-powered mill was sawing logs on the James River. Emigration companies looked for loggers and sawyers to settle the forested land.

The single central fact about the colonial American landscape was its forests. Nothing like these--hundreds of millions of acres of hardy softwood and hardwood forests--were to be found in modern Europe. These were woods fit for a king. And as long as the British Crown and British industries were willing to pay generous prices for pine ship masts from Massachusetts, potash from the mid-Atlantic hardwoods, and tar and turpentine from the Carolinas, the resourceful colonists were only too glad to help out. By the 1630s, logging and milling played as important a part in the New England economy as did fishing and the fur business. What could be better than to earn a shilling by cutting a tree? Commercial logging went hand in hand with settlement. The trees that at first seemed only a nuisance to farmers turned out to be a kind of green gold.

The way the Yankee traders put their green gold to use soon proved a source of conflict with England. The Americans were doing more than selling raw materials to England and building their own cargo and whaling fleets. Mercantile interests eclipsed political loyalties as the colonists sold hogsheads, firewood, and spar trees to England's enemies, Spain and Portugal. London shipbuilders, meanwhile, grumbled that they were losing their best shipwrights to New England. Local governments in New England infuriated the mother country by privatizing Crown land and refusing to let the king's agents cut trees on that land.

Few aspects of British rule were more galling to the Americans than were the "broad arrow" laws. These laws, aimed at supplying masts for the British navy, were named after the practice of marking the navy's spar trees with three slashes of the hatchet--the broad arrow. In a land so vast as America and so far from the seat of government, enforcement of the law proved a nightmare. A royal surveyor found in 1717 that only one of the seventy broad arrow pines in Exeter, New Hampshire, was still standing. The laws were rewritten more than once in an attempt to counteract the increasing rebelliousness of the colonies. Twice, in the 1720s, deputies sent by the royal surveyor general returned empty-handed from journeys to fetch poached lumber. In one case, the surveyor himself was threatened with death. In the other, citizens of Exeter dressed up as Indians--shades of the later Boston Tea Party--and beat the royal agents. When the agents beat a retreat to their boat, the sabotaged vessel sank and they were forced to walk back to Portsmouth.

The colonists' utilitarian views of the great American forest did much to shape their treatment of those woodlands. They saw in the woods a barrier to settlement. They saw a plentiful resource with which a nation could be built. But there was more to it than simple economics. The Europeans came to the New World from places where the landscape had been domesticated for hundreds and, in some cases, thousands of years. True wilderness was the stuff of ballads and epics. Here was something entirely new, and the settlers weren't sure what to make of it.

Having left behind the order of the artificial garden that was Europe, how were they to live in the seeming disorder of a wild land? In the family Bibles they brought with them, the settlers undoubtedly found special guidance in this injunction from the first chapter of *Genesis*: "Be fruitful and multiply, and replenish the earth and subdue it: and have dominion over the fish of the sea, and over the fowl of the air, and over every living thing that moveth upon the earth."

In the human-dominated world of the twentieth century, it is difficult to imagine the fear which many colonists felt upon their arrival in an untamed land. A typical account is that left by my great-great-great-great-grandfather John Ervin of his father's journey inland from Charles Towne, South Carolina:

> Oft have I heard my Father tell of his Pioneering enterprise--of how their small vessel crept Cautiously up the dark, tortuous reaches of the Black River, bordered w'h thickly-forested Swamps that shut out the Daylight. Their apprehensions for their safety increased when oft the Silence was shattered by Hideous & Unearthly screams of wild things. On reaching the Kings Tree, great was their surprise to find Nothing but Primeval wilderness. Notwithstanding the Company scattered to Select home-sites near Streams or Springs. The Irvines chose a Bluff about a Mile distant & set to work to Fell the mighty trees & Clear away undergrowth. A crude Shelter was erected tight on the sides of Prevailing winds. Later w'n joined by the Balance of the Family a large Cabin was Built with Thatched roof & mud Chimneys.

The wilderness was terrifying. At the most obvious level, the white settlers were concerned for their physical safety. What if they were to encounter a wolf, a panther, or a hostile Indian? What if they failed to build a home before winter set in? Or, perhaps most worrisome, what if they failed to hunt, gather, or grow enough food? But there was another sort of fear, too, a fear that their immortal souls were in danger. No one expressed this more graphically or more crudely than the Puritans of New England. More threatening than the arrows of the forest natives were their "pagan" beliefs. Michael Wigglesworth wrote of:

> . . . a waste and howling wilderness
> Where none inhabited
> But hellish fiends, and brutish men
> That Devils worshiped.

Cotton Mather, similarly, preached that Satan had established a stronghold among the natives. Indian leaders, he ranted, were "horrid Sorcerers, and hellish Conjurers and such as Conversed with Daemons."

The dark forest, inhabited as it was by "hellish fiends," provided a handy metaphor for the internal struggle that each believer must wage against the forces of evil. On a more literal level, such language put the church's imprimatur on a brutal campaign to subdue the heathens along with the forest. Mather held up the "glorious gospel-shine" as antidote to the temptations of the "forrests wide & great." Thomas Shepard saw a battle between "the cleare sunshine of the Gospell" and "thick antichristian darkness." This dualism of Gospel sunshine versus wilderness darkness was taken literally as well as metaphorically. The soldiers of Christ believed they were truly pushing back the forces of evil and darkness as they let light into the forest. The Puritans had gone well beyond the injunction of *Genesis*. God's chosen weren't just exercising dominion over every living thing; they were waging war against the wild things.

Those quotations from the Puritans are drawn from Roderick Nash's important study, *Wilderness and the American Mind*. In the preface to the revised edition, Nash considered the deepest roots of the settlers' antipathy to the forest. He argued that the answer lies in prehistory, some fifteen million years ago, when the ancestors of *Homo sapiens* responded to climatic changes and began to emerge from forest into grassland. In their new environment, these intelligent primates evolved superior vision, which helped them hunt animals that were larger and faster. In the jungle, the evolving human race was at a disadvantage to larger mammals blessed with superior senses of smell and hearing. "Thus once our ancestors left the wilderness," Nash wrote, "they were loath to return to an environment that neutralized their visual advantages. Indeed, when they could they burned forests in order to convert them to open grassland." He attributes our preference for view home sites and children's fear of the dark to a lingering sense that we are less secure in the forest.

The idea that mankind must beat down a recalcitrant natural world has not died. A timber-industry executive put it this way a few years ago: "We have the directive from God: Have dominion over the earth, replenish it, and subdue it. God has not given us these resources so we can merely watch their ecological changes occur."

At first, the deforestation proceeded slowly. Despite the conflicts with Britain and the growing trade in forest products, the forests of North America were felled primarily to make way for farms and settlements. Axe-wielding settlers could clear the land only so fast. A homesteader was likely to spend a whole month cutting down one acre of old-growth trees. More time was spent removing logs, rocks, and--if the farmer was ambitious--the huge stumps. Although the fallen trees put roofs over the settlers' heads and fire in their hearths, the woods often were seen more as an inconvenience than a blessing. The slow, back-breaking process of subduing the forest helped shape the early nation. Richard G. Lillard observes in *The Great Forest*:

> As the woods thus stood against the spread of settlement, communities grew up strong and thickly populated. The same had happened in the natural clearings of southern England in Roman times and in openings in Peru in the great days of the Incas. Where

forests hem in a dynamic culture, it integrates and matures. In 1700 American colonists could ride from Falmouth [now Portland], Maine, to Williamsburg, Virginia, sleeping each night in a considerable village.

For every acre of land cleared to supply sawmills or to expand cities, thirteen others were cut to make way for farms. Most of the fallen trees were burned, either as logs in fireplaces or as charcoal in small industrial furnaces. It was an axiom of American settlement that the plow followed the axe. This held true until the Civil War era. By then the basic facts of American life changed. As the United States rapidly industrialized, the nature of deforestation changed. No longer was agricultural expansion what drove logging. Now trees were being felled primarily to produce the lumber that would house millions of immigrants and build the railroads. A new breed of migrant loggers was cutting the timber. In the wink of an eye, the American landscape was altered. More than 100 million acres of forest were cleared between 1850 and 1879--roughly equivalent to the amount cut during the previous two and a half centuries.

Not only did the pace of logging intensify in the latter half of the nineteenth century, it underlay the westward march that would continue well into the twentieth century. New York gave way to Pennsylvania as the leading timber producer. Then the lumbermen discovered the pine forests of the Great Lake states. By 1872, Michigan was cutting more timber than New York and Pennsylvania combined. Michigan was being stripped of its woods at a pace never seen on the face of this planet. The era of high-volume, cut-and-run logging was in full swing. When the loggers finished with Michigan, the land was left as bare as the scalp of a Marine recruit at boot camp. As often as not, mill owners abandoned their timberland, forfeiting it for unpaid taxes. (In many cases they had no land to forfeit, having helped themselves to timber on the public lands.) No thought was given to reforestation. There was always more virgin timber farther west or farther south or farther north.

State after state was denuded of its trees. Michigan was followed by Wisconsin and then Minnesota as the leading lumber producer. When the end was in sight for cheap timber in the Lake states, the increasingly rich and powerful lumber barons looked to the South and the Pacific Coast for their future supplies. In a few generations, the new Americans had done what it took thousands of years for the Europeans to do and what the native Americans had never done: deforest a continent. They had taken to heart, with a vengeance, Thomas Jefferson's observation that this enormous land mass contained "great quantities of land to waste as we please."

Jefferson never dreamed, in his agrarian day and his agrarian world view, that whole forests from Maine to the Louisiana territory would be laid waste to feed the insatiable appetite of industry and metropolis. It has been said that in precolonial America a squirrel could have made its way from Maine to

Louisiana by leaping from the branches of one giant tree to the next. Three and a half centuries after the first English colonists went ashore at Jamestown, only a few patches of old-growth forest remained east of the Great Plains. In the West, the low-elevation forest was heavily cut and the rest was going fast. An incomprehensibly huge forest, one that had survived many millennia of human occupation, was rapidly subdued.

The self-sufficient yeoman of Jefferson's dreams had given way to the wage laborer of the modern world. No longer did the plow follow the axe. Early in this century, a back-to-the-land movement briefly flourished, with the goal of persuading settlers to farm the logged-over lands of the West. A few hardy souls tilled the infertile, erosion-prone forest soil; most gave up quickly. In the axe's wake, there was nothing to do but grow another forest for the axe.

Few Americans in the nineteenth century voiced the idea that there might be something seriously wrong with a Manifest Destiny that meant subduing the natural world along with its first human inhabitants. One of the most eloquent protests came from an elder of Puget Sound's Duwamish and Suquamish peoples. Chief Seattle's prophetic warning--undoubtedly embellished by his translator--was contained in a letter to President Franklin Pierce in 1854, the year the tribes were forced to sign treaties with the U.S. government:

> We know that the White Man does not understand our ways. One portion of the land is the same to him as the next, for he is a stranger who comes in the night and takes from the land whatever he needs. The earth is not his brother, but his enemy, and when he has conquered it, he moves on. He leaves his father's graves, and his children's birthright is forgotten. The sight of your cities pains the eyes of the Red Man. . . .

> . . . The Indian prefers the soft sound of the wind darting over the face of the pond, the smell of the wind itself cleansed by a mid-day rain, or scented with a piñon pine. The air is precious to the Red Man. For all things share the same breath--the beasts, the trees, the man. . . .

> What is man without the beasts? If all the beasts were gone, men would die from great loneliness of spirit, for whatever happens to the beasts also happens to man. All things are connected. Whatever befalls the earth befalls the sons of the earth.

> It matters little where we pass the rest of our days; they are not many. A few more hours, a few more winters, and none of the children of the great tribes that once lived on this earth, or that roamed in small bands in the woods, will be left to mourn the graves of the people once as powerful and hopeful as yours.

The whites, too, shall pass--perhaps sooner than the other tribes. Continue to contaminate your bed, and you will one night suffocate in your own waste. When the buffalo are all slaughtered, the wild horses all tamed, the secret corners of the forest heavy with the scent of many men, and the view of the ripe hills blotted by the talking wires, where is the thicket? Gone. Where is the eagle? Gone. And what is it to say goodbye to the swift and the hunt, the end of living and the beginning of survival? We might understand if we knew what it was that the White Man dreams, what he describes to his children on the long winter nights, what visions he burns into their minds, so they will wish for tomorrow. But we are savages. The White Man's dreams are hidden from us.

By the time English and American explorers took a close look at the Pacific Northwest coast, they were beyond the extreme fears expressed by the Puritans in New England. They still would describe the Indians as "filthy," "nasty," and "indolent." But the forest of tall conifers made visions of pounds and dollars dance in their heads. The industrial revolution was getting underway, and they knew the value of plentiful timber. What they found on the coast of Oregon, Washington, and California was the greatest timber-growing land in the world. The forests of the Douglas fir region typically held five times as much timber as an acre of old growth in the East.

When British explorer George Vancouver in 1792 unwittingly sailed past the mouth of the Columbia River, the "luxuriant landscape" drew his attention: "The whole had the appearance of a continued forest extending as far north as the eye could reach, which made me very solicitous to find a port in the vicinity of a country presenting so delightful a prospect of fertility." When Vancouver's ship, the *Discovery*, needed a new fore-topsail yard, the captain proclaimed his luck at being able to "make our choice from amongst thousands of the finest spars the world produces."

Meriwether Lewis, who reached the coast by land a decade later, seemed more interested than either Vancouver or Vancouver's botanist Archibald Menzies in the trees themselves. In the great forest along the lower Columbia, Lewis could barely contain his excitement over a fir tree that had a forty-two-foot circumference and shot two hundred feet straight up before its lowest limb. This tree, he wrote, was "perfectly sound." And he found from other specimens that its wood "rives better than any other species." Botanist David Douglas, after whom the Douglas fir came to be named, called it "one of the most striking and truly graceful objects in Nature."

The enthusiasm expressed by the explorers and botanists was modest compared to that felt by the first loggers. For more than two centuries they had worked their way across the continent. Then, in the far corner, they found the best timber to be found anywhere. Stewart H. Holbrook, the premier chronicler of American logging history, has the boys practically jumping out of their calks in amazement:

Left: Loggers who journeyed from the Midwest to the pacific Coast couldn't believe their eyes when they saw trees like these Sitka spruce in Lincoln County, Oregon. When they got to work, the forest fell fast. Below: Railroads opened up millions of acres of logging and remained the standard means of transporting logs until the 1940s.

When the first loggers saw the fir that grew along the banks of the Columbia and around Puget Sound they said there couldn't be timber that big and tall. It took, so they told each other, two men and a boy to look to the top of one of these giants.

And thick? Holy Old Mackinaw, the great trunks stood so close that the boys wondered how a tree could be felled at all! And between the trunks grew a jungle of lush growth that no Maine or Michigan logger had ever imagined. You actually had to swamp out a path to a tree and to clear a space around it before there was room to swing an ax. . . . It would take some doing, mister, to let any daylight into *this* swamp.

The mill owners and the rough breed of men who worked for them moved across the continent in two waves. It wasn't until the second wave that they reached the coast. First they migrated westward from Bangor, Maine, where they had built the greatest concentration of sawmills ever seen. Many, like lumberman Thomas Merrill, made a beeline from Maine's Penobscot River to the new timber frontier in the woods around Saginaw, Michigan. The loggers continued to push the frontier across the Lake states until the end of that forest was in sight. Then they made for the Pacific Coast and the tallest trees of all. A few, like Andrew Pope and William Talbot, didn't bother to stop off at the Great Lakes. From East Machias, Maine, they made their way to Washington, where their Puget Mill Company built sawmills on both sides of the Hood Canal. Catering primarily to the lucrative California market, their sawmills and their shipping fleet comprised the largest lumber operation in the Northwest from the 1850s into the early 1900s.

But the dominance of the transplanted New Englanders who settled around Puget Sound and Hood Canal wasn't to last long. In the 1880s, San Franciscan A.M. Simpson brought high-volume lumber production to Grays Harbor, Washington--a port snuggled into spruce- and fir-clad hills that may have held the very richest woods of the Northwest. He was quickly followed by the likes of Cy Blackwell, who came by way of Maine, Minnesota, Wisconsin, and California. A few Canadians--Sol Simpson from Quebec and Alex and Robert Polson of Nova Scotia among them--would make their way to Douglas fir country as well. But in the end, the biggest landholder and mill owner of them all proved to be another Lake states lumber baron: Frederick Weyerhaeuser.

Known to his employees as "Dutch Fred" because of the thick accent he brought with him to America, the young Weyerhaeuser had gotten his start as manager of a sawmill in Rock Island, Illinois. In partnership with his brother-in-law Frederick Denkmann, he bought the mill when the financial panic of 1857 threw its owners into financial difficulties. Under its aggressive new ownership, the mill was tremendously successful. Denkmann ran the shop while Weyerhaeuser handled the business of buying logs and selling lumber. It wasn't long, though, before the partners took the first step toward putting together the largest privately owned timberland empire in the world. Weyerhaeuser, determined to shake his reliance on independent loggers, began acquiring timberland in the pine forests of Wisconsin's Chippewa River region. He kept buying land and mills until companies under his control dominated the industry in six states. By the end of the century, he controlled twenty lumber and railroad companies, managing his affairs from the German-American Bank of St. Paul, Minnesota, known popularly as "the Weyerhaeuser bank." His biggest deal was yet to come.

After moving to St. Paul, Weyerhaeuser began to do business with his next-door neighbor on Summit Street. The neighbor was railroad tycoon James J. Hill, who sold pine-covered land in Wisconsin and Minnesota to the Weyerhaeuser interests. Both men benefited from the largesse of a federal government eager to encourage the westward extension of rail lines. The Northern Pacific Railroad--of which Hill later took control--was granted twenty square miles of land for every mile of track it laid between Duluth, Minnesota, and Portland, Oregon. In all, the Northern Pacific received a 39-million-acre land grant. Some of the most valuable land was old-growth forest on the west side of the Cascade Mountains. And who could make better use of the timber than Hill's friend and neighbor, Frederick Weyerhaeuser?

The value of the property had been boosted considerably by Congress' willingness to write a most accommodating provision into the Organic Administration Act of 1897. That provision allowed trades of private land within the new forest reserves for other public land. Hill used the law to exchange worthless land covered by rocks and glaciers around Mount Rainier for coastal property with gigantic Douglas firs. The federal deal done, it was up to Weyerhaeuser and his fellow investors to come to terms with Hill. "Dutch Fred" and several business partners traveled to southwest Washington in November 1899 to take a look at timberlands and mills. After some dickering with Hill, the Weyerhaeuser interests bought 900,000 acres from the Northern Pacific at $6 an acre--a deal that electrified the timber industry. In one bold stroke, the new Weyerhaeuser Timber Company became the second largest timber landowner in the United States.

Some observers thought Weyerhaeuser paid a lot for the land and stumpage. But one historian has noted the $5.4 million deal worked out to 30 cents per thousand board feet of old-growth timber. (A board foot is a measure equal to a piece of wood one foot square and one inch thick.) In a few short years, the value of the wood rose to $2.50. Today, Weyerhaeuser is logging the last of its old growth and is exporting much of it as raw logs for hundreds of dollars per thousand board feet. Weyerhaeuser Timber did not immediately begin logging its lands in a big way. Instead, the company manager, George S. Long, promptly moved from Wisconsin to Tacoma and began adding to Weyerhaeuser's real estate portfolio. The company bought more land from the Northern Pacific, southern Oregon property from the Southern Pacific Railroad, and other virgin timberland from other owners. Today, Weyerhaeuser is the largest timber landowner in the world, with six million acres in the United States and long-term leases on more than seven million acres of Crown land in Canada.

Frederick Weyerhaeuser's bold and sudden move to the coast served notice that the timber industry's center of gravity was shifting rapidly to the Pacific Northwest. And that the big investors from the Great Lakes, backed by big capital, would play the tune to which others would dance. The industry already had been moving into wealthier and wealthier hands. The big trees close enough to the mills to be hauled by ox or horse teams down a skid road had already been cut. So, too, had the woods adjacent to waterways large enough to float the logs to mills. From now on, timber would be cut by those with enough capital to lay railroad track and build trestles into the woods.

Cut the timber they did. One thirty-six-square-mile township on the southwest corner of Washington's Olympic Peninsula produced an immense amount of wood. The big cut began in 1909, three years before timber cruisers for Chehalis County walked through "Twenty-one-nine" (Township 21 North, Range 9 West) to inventory the merchantable timber for tax purposes. The Douglas fir were so closely packed and so thick (up to fourteen feet in diameter) that it proved difficult to count the trees, much less estimate the volume of timber. The county government socked landowners with heavy property taxes on the woodlands, and the owners responded by leveling the woods as fast as they could. The harder they worked, the lower their taxes. By the time the stock market crashed in 1929, Washington had become the nation's top lumber producer. Much of that cut was from Grays Harbor, where at least two dozen sawmills were working around the clock, seven days a week, to saw the big trees from Twenty-one-nine and the townships around it. Grays Harbor shipped out more lumber in those days than any other port before or since.

As the logging business fell more and more into the hands of corporate interests, the men in the woods grew increasingly alienated from the owners. The stage was set for bitter and often violent labor disputes.

But as long as loggers lived in remote camps, their rough-and-tumble lifestyle didn't change much. While they were taming the wilderness, they remained untamed themselves. Unless they quit in disgust over camp food or working conditions, loggers typically went to town only twice a year: the Fourth of July and Christmas. Saloons and brothels in the towns and cities of timber country were ready for those occasions. The drinking, whoring ,and fighting of the early loggers were legendary. Many of these rough men carried "logger's smallpox" on their skin: the marks left during a fight by the spikes of another logger's calked boots.

Portland's most successful lumberman, Simon Benson is said to have found a clever way to keep his logging camps operating during the busy summer months. Most companies lost money after their crews went to town for the Fourth of July, returning to camp only after a bender that frequently lasted three weeks. Supposedly, a barrel of whisky would appear at the Benson camps around the middle of June. The Scandinavian loggers would drink their fill. Then, convinced that they had had their Fourth of July fun, they would be hard at work while Benson's competitors were shut down.

"Loggers has always been quite a drinking tribe," retired logger Dewey Bryson of Darrington, Washington, told me. "I went to one doctor in Concrete. He said, 'You know why you guys are in such good health? It's because you guys get so drunk Friday night and puke all the poison out of yourselves by Monday.'" As a child, Bryson lived for several years on a railroad car outfitted as family housing in a logging camp. After he became a logger himself, he lived on the family homestead and commuted to work by rail on a "speeder" car.

The notorious excesses of the prototypical logger may have had as much to do with the risks of living on the edge as with the loneliness of living in all-male bunkhouses. Logger-turned-author Stewart Holbrook recalls the day during his youth in turn-of-the-century New Hampshire when the whole town of North Stratford turned out to watch the loggers break up a massive logjam on the Connecticut River. Slim Pete Hurd, a hero of the local kids, got his peavey on the "king log." As happened so often, the logs broke loose with a sudden and violent crash of white water. "Pete's gone!" someone shouted. The townsfolk walked back home slowly and silently. Another fearless man had died.

Logging wasn't any safer on the West Coast, particularly during "highball" years when bosses pushed workers to work faster and faster. One hundred men lost their lives while highballing in the woods of Grays Harbor in 1925. All logging jobs are dangerous. But probably none surpassed the spectacularly risky business of high rigging. When high-lead logging first came in, a man would climb a sturdy tree, limbing it on the way up, then top it and rig it with all the cables needed to haul logs and hold the spar tree steady. A steam "donkey"--the machine that replaced the last horse teams--would pull logs to the landings. The last of the spar trees were replaced by steel towers in the 1970s. The legends of high riggers live on.

Haywire Tom Newton was the king of the high riggers. His exploits were far more impressive than those of Paul Bunyan because Newton was a flesh-and-blood mortal. It is possible, of course, that a slight bit of exaggeration crept into some of the stories told about Haywire Tom. Newton refused to wear a safety belt when he climbed the tall trees. As the top crashed to the ground and the spar shook violently, he would jump to the cut-off surface and ride the tree with his arms outstretched. Once he slipped and fell 120 feet. Lady Luck was still with him. He landed feet-first in deep mud, able to smile and wave to his co-workers as he was loaded into an ambulance. After an examination at the hospital, he went right back to the logging camp. Newton died years later when a crane tipped over while he worked on the Grand Coulee Dam.

Top: The chain saw, like the one being used above to fell an Alaska yellow cedar in the Washington Cascades, has replaced the two-man "misery whip" of earlier years. Left: But the worker splitting cedar into shakes in Darrington, Washington, is using much machinery similar to that used for generations.

Much has changed in the woods since the 1930s, when loggers lived in camps, cut trees with the two-man crosscut saws known as "misery whips," loaded logs onto trains, and were the undisputed heroes of the children of the Northwest. The camps are all gone, grown over with weeds or Douglas fir. The misery whip and railroad logging gave way in the 1940s to the chain saw and the logging truck. More wood is being cut today by fewer men. (Women are still virtually unheard of in logging crews.)

Loggers are still cutting old growth, though. Of course, there's no more old growth to be cut on the East Coast. Except for a few forest stands protected in parks, no uncut natural forest remains east of the Mississippi River. When logging began on this continent, nearly a billion acres of natural forest covered what are now the lower forty-eight states. The great bulk of those acres lay east of the Mississippi. Given the immensity of those woods, it's remarkable that Americans were able to log off the entire eastern forest so quickly.

What's left of the original American forest lies along the Pacific Coast or straddles the Rocky Mountains. Even there, the most valuable timber already has been cut. What's left is almost entirely on federal land on the steep slopes of the Cascade Mountains. Everything outside the national parks and wilderness areas is up for grabs. Roughly two-thirds of the remaining old growth in Oregon and Washington is unprotected. As long as merchantable trees remain standing, there are those who will want to cut them. Because the old trees are "shot with decay," as one forester put it, and because there is a market for the clear, tight grain of old-growth wood, the timber industry feels the trees should be cut. After all, as former Forest Service Chief William B. Greeley wrote, "Forestry begins with the ax."

Early in this century, the chief of the Forest Service, Gifford Pinchot, ordered a survey of loggers that asked them numerous details of their business. The final question had to do with reforestation: "What provision do you make for reproduction?" William B. Greeley, who was Pinchot's assistant at the time, recalled the answers to this question:

> Most of those questioned left this line blank. One operator in Minnesota wrote in: "Nothing of the kind allowed in my camps." As far as renewing trees was concerned, this reply fairly summarized the situation. . . .
>
> Even in those days the industry was still dominated by "free timber." Its shadow hung over the lumber markets from coast to coast. Any sawmill in the West could pick up incomparable old-growth stumpage in the next valley or the next county for four bits or a dollar per thousand feet. Men conserve things that are scarce and valuable; and notwithstanding the prophecies of a

shortage, timber remained a cheap commodity by the cold standards of commerce.

Occasionally, industrialists' reliance on this continent's forested plenty turned out to be an economic mistake. For two centuries, American whalers and cargo ships were the envy of the world. The U.S. merchant marine, consisting largely of sturdy wooden "down-easters," was the world's biggest and best by 1850. But advantage turned to disadvantage in the following decades. American shipbuilders continued to produce wooden sailing ships while the rest of the world began building metal-hulled steamers. Before long, this failure to keep pace with technology foorced Americans to ship their goods on foreign-flag vessels. The timber barons, similarly, gave no more thought to growing another crop of trees than New England's ship builders gave to the looming competition from metal-working shipyards. Their failure--and ours--to think about the forests of the future is remarkable in light of the fact that forests have been a major source of this nation's wealth for nearly four centuries.

But, as Greeley noted, the sheer abundance of this continent's forests led to the mistaken notion that this source of wealth was "free." During the great westward migration, the government distributed hundreds of millions of acres of forestland. Much of it was given away free, much of it sold for a song. Law after law was passed for the purpose of distributing public land to the homesteaders who would form the backbone of the economy. Time and again, that fell into the hands of lumbermen. Sometimes state governments acted as intermediaries for this growing corporate control. The Swamp Land Act of 1850, for instance, gave to the states 63 million acres that were to be distributed to settlers who would drain the lands and farm them. Michigan sold off its 5.6 million acres of "swamp" (much of which contained valuable timber) for $1.25 an acre, no questions asked. Nearly a million acres went to an iron company on the Upper Peninsula. Florida sold off four million acres to a single buyer--hardly a penniless homesteader. So much for land reclamation.

The federal government's failure to guard its land resulted in the infamous "Round Forty." A lumberman would buy a forty-acre parcel, set up a logging camp there, and proceed to log everything within reach--most of it on the adjacent public land. The Timber and Stone Act of 1878 allowed settlers to buy up to 160 acres of public land for a modest $2.50 an acre. When constituents in logging states complained that the law imposed a hardship, Congress accommodated them by lowering the price to $1.25 and allowing loggers to pay for the land *after* logging it.

Most of the federal land acts were intended to benefit individuals buying a single parcel, not lumbermen cutting whole sections and townships. Logging companies in the Great Lakes states and the Douglas fir region evaded the law by hiring crews of "entrymen" to march into federal land agents' offices, buy homesteads, and subsequently transfer the land to the silent partners. Some entrepreneurs dispensed with the legal niceties altogether and simply helped themselves to publicly owned timber without any pretense of paying for it. In the rare instances when federal prosecutors tried to take timber rustlers to court, judges and juries excused the defendants as just following local custom. To this day, many residents of timber towns wink at log thefts from the national forests. The only difference is that today's thefts usually involve a few high-value red cedar logs, not whole watersheds.

Those abuses of the public lands, coupled with a growing concern about the future timber supply, led to creation of the national forests. Congress passed the Forest Reserve Act in 1891, authorizing the president to designate portions of the public domain as "forest reserves." Almost immediately, President Benjamin Harrison established the first reserves. Over the following century, the national forests, as they came to be known, have grown to 191 million acres. Contrary to widespread misconception, the national forests are not part of the national park system administered by the Interior Department. Rather, they are managed by the Department of Agriculture under a "multiple-use" mandate intended to accommodate everything from clearcutting and tree farming to unspoiled wilderness.

Gifford Pinchot, first chief of the Forest Service and in many ways still its guiding light, was obsessed with the idea that the nation would suffer a "timber famine" sometime in this century. The solution he preached was the kind of sustained-yield forestry he had learned in Europe. Pinchot was the first to popularize the term "conservation," warning that disaster awaited if trees continued to be cut without any thought of tomorrow. Before Pinchot, nobody had gone any further in the direction of forest conservation than had the founder of Pennsylvania, William Penn. This English Quaker insisted that settlers leave one acre of land forested for every four acres cleared. Well into this century, waste was the norm. "You would just take the merchantable trees," recalls Charlie White, who began logging in 1928 and is now mayor of Darrington, Washington. "You wouldn't take the ones yea big [holding his hands a foot and a half apart]. The mills wouldn't buy them." The slash left on the ground by many Pacific Northwest logging crews exceeded the total amount of timber that could be pulled out of the woods in the East.

Pinchot's idea of conservation was radically different than that of his sometime friend, sometime foe, John Muir. Pinchot's mission was to teach America how it could produce a never-ending supply of timber. Muir's goal was to teach Americans to love the wilderness for its own sake--and to preserve it. The thoroughgoing nature of his devotion to the natural world was something new to Pinchot. The two men were hiking in the Grand Canyon when Pinchot prepared to shoot a tarantula. "He wouldn't let me kill it," the astounded Forest Service chief later wrote. "He said it had as much right there as we did." Few humans have written of the old-growth forests with as great a passion or as much personal knowledge as did Muir, founder of the Sierra Club. To him, the big woods of California and the Northwest were "temples," the trees "psalm-singing." When he described loggers felling a tree in western Washington, it was as though he were watching a person fall: "[T]he noble giant that had stood erect in glorious strength and beauty century after century, bows low at last and with gasp and groan and booming throb falls to earth."

A third perspective that helped fuel the growing consensus that America's forests should not be turned into a coast-to-coast clearcut was that of George Perkins Marsh. Marsh did not concern himself about the possibility that sawmills might face future shortages. Nor was the human soul's relation to the natural world the issue for him. Marsh's contribution was to document how one Old World civilization after another had destroyed the fertility of its land through deforestation and careless soil management. The influence of his rigorously researched work on twentieth-century thought can scarcely be exaggerated. Published almost a full century before Rachel Carson's *Silent Spring*, Marsh's *Man and Nature* was the most important work of its time to explore humankind's ability to alter the natural environment. The experience of the Mediterranean cultures offered valuable lessons for the New World:

> With the extirpation of the forest, all is changed. At one season, the earth parts with its warmth by radiation to an open sky--receives, at another, an immoderate heat from the unobstructed rays of the sun. Hence the climate becomes excessive, and the soil is alternately parched by the fervors of summer, and seared by the rigors of winter. Bleak winds sweep unresisted over its surface, drift away the snow that sheltered it from the frost, and dry up its scanty moisture. The precipitation becomes as irregular as the temperature. . . . The soil is bared of its covering of leaves, broken and loosened by the plough, deprived of the fibrous rootlets which held it together, dried and pulverized by sun and wind, and at last exhausted by new combinations. . . . The earth, stripped of its vegetable glebe, grows less and less productive, and, consequently, less able to protect itself by weaving a new network of roots to bind its particles together . . . .

In most countries of Europe--and I fear in many parts of the United States--the woods are already so nearly extirpated, that the mere protection of those which now exist is by no means an adequate security against a great increase of the evils which have already resulted from the diminution of them.

It is startling that an architect-turned-diplomat would be the author of such an important scientific work. Subsequent studies by scientists with all the appropriate credentials have confirmed this Renaissance man's finding that deforestation, intensive agriculture, and overgrazing were responsible for reducing much of the Mediterranean region to desert. Marsh was careful to point out that most of North America is not as vulnerable as the Mediterranean landscape to erosion and desertification. It's important also to note that present-day logging in the Douglas fir region and many other areas generally is followed by reforestation, not by cultivation of grains or grazing of cattle. Yet there's no escaping the fact that erosion is taking a severe toll on many of America's forestlands and croplands. A few once-rich forests, such as Tennessee's Copper Basin, have indeed been turned into deserts. Repeated fires have left New Jersey's pine barrens incapable of growing the tall trees that once stood there.

One can only wonder what Marsh would say about the destruction of natural forests around the world in the face of the major climatic changes now underway. The rain forests of the tropics and the temperate rain forests of North America both continue to be felled and the burning of their slash adds to the atmospheric burden of carbon. As I write these lines, a thick layer of smoke from slash burning on the Olympic Peninsula hangs over Seattle. The tropical deforestation of the late twentieth century recalls Marsh's Mediterranean. Clearings are made in the forest so that settlers can plant crops, but the jungle soils are too thin and poor to support the crops. So cattle are grazed on the delicate soil until it all washes away--in a short period of time. Without trees to transpire water into the air, rainfall decreases. A desert is created. Even though this planet's forest reservoirs of carbon represent an important hedge against accelerated greenhouse warming, fifty acres of tropical jungle disappear every minute. The old-growth forests of western Oregon and Washington are going somewhat slower--perhaps 190 acres a day--but unlike deforestation in Brazil or Indonesia, this alteration of the landscape is a direct matter of U.S. government policy. Our tax dollars make it possible.

In the face of global changes that could mean the inundation of coastal cities and desertification of this nation's breadbasket, it may seem a luxury to worry about the internal workings of the forest ecosystem. But in those forests, plant species of use to humans are waiting to be discovered before they become rare or are driven to extinction.

Miracle drugs or no, greenhouse antidote or no, more concerned citizens are coming to the defense of ancient forests as an ecosystem of value in itself. No one seriously expects that the spotted owl will ever have much *practical* value for humans. Yet today's environmentalists believe passionately, as did Muir before them, that this ecosystem and its inhabitants have as much right to exist as we do. It's a perspective that is filtering down to the children of this land. Dr. Seuss seemed to have the ancient forests in mind when he wrote his environmental masterpiece, *The Lorax.* When the protagonist, the Once-ler, tells how he came to a beautiful land and saw the trees, it is reminiscent of what the Michigan loggers saw when they set eyes on the Northwest's tall timber:

> And I first saw the trees!
> The Truffula Trees!
> The bright-colored tufts of the Truffula Trees!
> Mile after mile in the fresh morning breeze.

Businessman that he is, the Once-ler soon is busy cutting down Truffulas and turning their tufts into Thneeds ("A Thneed's a Fine-Something-That-All-People-Need!"). The business grows by leaps and bounds, disturbed only by an annoying little creature named the Lorax, who speaks for the trees. And he speaks for the animals: the Brown Bar-ba-loots who eat Truffula Fruits, the Swomee Swans who can't sing with smog in their throats, and the Humming-Fish whose gills become gummed with Gluppity-Glupp. The Once-ler is informing the Lorax that he has his rights, too, when the inevitable occurs:

> And at that very moment, we heard a loud whack!
> From outside in the fields came a sickening smack
> of an axe on a tree. Then we heard the tree fall.
> *The very last Truffula Tree of them all!*

The Thneed factory shuts down and the Lorax disappears. But there's a ray of hope. The Lorax hands to a boy the last Truffula Seed. The Once-ler, guilty over creating a wasteland, tells the boy:

> Plant a new Truffula. Treat it with care.
> Give it clean water. And feed it fresh air.
> Grow a forest. Protect it from axes that hack.
> Then the Lorax
> and all of his friends
> may come back.

"May" is the key word. If the Bar-ba-loots' habitat is destroyed, all the captive breeding and reforestation in the world won't save them. Seuss's story, written in 1971, reads as a parable for the struggle over the virgin forests. We have no Lorax who can tell us the role that every plant, animal and mineral plays in the old-growth ecosystem. We have to learn that for ourselves. Even as the clock ticks away on the ancient forests, scientists are working to crack its mysteries.

4

Thinking Like a Forest

The darkness is almost total. I've just shut the door of my car when I hear the engine and see the headlights of a vehicle coming up the hill. That car pulls over to the side of the dirt road as I watch warily, not entirely comfortable about encountering strangers in the dark so far from the nearest town. I can barely make out the figure stepping from the driver's side and walking toward me. It's a woman on the later side of middle age, well bundled against the predawn chill. She's carrying a flashlight and a notebook. We introduce ourselves and discover we've both come here for the same reason.

In this clearcut carved out of the old-growth forest we're going to look and listen for the marbled murrelet, an elusive robin-sized seabird. Jean Cross tells me she's been visiting this clearing nearly every morning for the past month. As we talk, the stars slowly fade and the forest's edge is silhouetted against a pale sky. Two more cars pull over to join this birdwatching spectacle. Out of one car steps a young Forest Service biologist and a Washington Wildlife Department volunteer, out of the other two National Audubon Society members.

It's just after five in the morning when we hear the first sounds of murrelets. *Aheer! Aheer!* they shout in high-pitched voices. Their piercing, gull-like cry would be perfectly in place in a harbor or an oceanside beach. But we are standing on a hillside above the South Fork Stillaguamish River, more than twenty miles from salt water. This valley, dotted with ghost towns from gold-mining days, is located in the Mount Baker-Snoqualmie National Forest northeast of Seattle.

During the hour and a half we spend together, we hear the cry of murrelets many times. When we first hear a bird's call, it comes from the direction of the trees fifty yards south of us. Seconds later, the cries are repeated above our heads or to the north. Some birds fly by to our right, others to our left. Sometimes the cries seem to come from one bird, other times from a group.

The sounds tell us the murrelets are headed toward the northwest, down the Stillaguamish valley. We scan the sky with our unaided eyes and our binoculars. Most of us don't see any birds except for a Cooper's hawk headed into the old growth with some prey in its talons. The Forest Service biologist wonders if she's just seen some murrelets. "I saw some black dots against the ridgeline," she says.

"That's all you'll see," Cross responds, confirming the sighting.

The South Fork Stillaguamish is the only Puget Sound drainage in which these birds are known to nest. Murrelets also have been seen along Canyon Creek, which spills into the Stilly a few miles downstream from here. And yesterday, Cross tells us excitedly, she was farther up the South Fork drainage near Gordon Creek when a low-flying murrelet, its small wings beating almost as fast as a hummingbird's, passed within twenty feet of her.

She's still trying to make sense of what she observed last week. As usual, the birds would fly purposefully over this clearcut on their way to salt water. But for a few days, other murrelets would come out of the woods, circle around for a while and then disappear. "I call them the kids, though I don't know for sure," Cross says of the second group. She believes they were fledglings testing their wings. That circling behavior has stopped, suggesting that the young now join the adults for the daily journey downriver.

Cross and other volunteers are helping biologists from the Washington Department of Wildlife study the extent to which the marbled murrelet uses old-growth forests. The discovery that this seabird frequents forests at all--and particularly stands like this one, so far from the shore--has generated considerable excitement among avian biologists. The marbled murrelet, named after the mottled brown-and-gray summer coloring of its nether side, has long been known as a bird that frequents saltwater shores. It fishes in the surf for sand lance, herring, and invertebrates. For many decades, scientists wondered where the bird nested. All that was known about its nesting habits was that it didn't nest with other shore birds on seaside rocks or cliffs.

Clues about the murrelet's nesting habits slowly accumulated. One observer in 1909 was thirty-five miles up the North Fork Nooksack River when he heard "an invisible party" of murrelets apparently following the valley toward Bellingham Bay. Over the next six decades, adult or juvenile murrelets were occasionally sighted in the ancient forests of Washington, Oregon, California, and British Columbia. But look as they might, biologists could not find a nest. The title of a 1934 article, "The Mystery of the Marbled Murrelet Deepens," conveys their perplexity. Then came two revealing incidents in British Columbia. In one, a logger and amateur bird watcher on the Queen Charlotte Islands found a stunned murrelet and an egg in the debris of a large hemlock that had been felled. On Vancouver Island, two murrelets plunged from a cedar as it was being cut down.

A marbled murrelet nest was finally found in Siberia in 1961. The nest was in a larch on a wide flat surface formed by a network of lichen-covered branches and twigs. Evidence that the North American variety of the bird also nests in trees came in 1974. Tree surgeon Hoyt Foster was at work in Big Basin Redwoods State Park in California's Santa Cruz Mountains when he found a downy chick in the crown of an old-growth Douglas fir. The nest, apparently used for years, was simply a depression in the moss on a flat limb high above the ground. A 1976 survey of Big Basin Park produced an estimate that one hundred breeding pairs nest there. Biologists apparently have solved the "mystery of the marbled murrelet." During the nesting season at least, the bird forages on the water by day and returns to its nest in old-growth or mature conifer forest by night. At day's end, groups of murrelets are sometimes seen gathering for their homeward trip. An exception to this nesting behavior is the Alaskan tundra, where the murrelet nests on the ground or on rocks.

A relative of the marbled murrelet also has an affinity for late-succession forests. The ancient murrelet nests in old-growth forests of hemlock, cedar, and spruce on British Columbia's Queen Charlotte Islands. These birds, instead of nesting on branches, burrow into the forest floor.

Solving the initial mystery of the marbled murrelet's nesting habits has raised new and troubling questions. With most of the coastal old growth already gone, has the murrelet population been reduced to a perilously small number? Can the bird make do in second-growth forest? Does it nest in alpine meadows as well as in old-growth trees? If the marbled murrelet can reproduce only in older forests, how much forest does it need? Where should habitat areas be located? Must federal and state timber sales programs be modified to protect this species? So far, there are no answers to those questions. Some nesting sites are protected already, for instance the Douglas fir-redwood forests of Big Basin Park and the cedar-hemlock groves of the research natural area, where the murrelets I heard are believed to nest. In other suspected nest areas, Canyon and Gordon creeks among them, logging continues.

Barry Troutman, who launched Washington state's nascent murrelet survey, likens the search for nests to "looking for a real small needle in a real big haystack." With years of work needed to define more precisely the bird's habitat needs, he worries that scientists will be expected to prove the bird is in immediate danger before any protective measures will be taken.

"Right now," Troutman says, "there appears to be a very strong association between these birds' nesting habitat and older forests. We have every reason to believe that's what's going on, and if nothing is done until the day you can prove beyond any shadow of a doubt that old growth is required, there's going to be precious little old growth. The chain saws are running." The biologist adds that these are his own views and not necessarily those of the Department of Wildlife.

Under contract to the Portland Audubon Society, consulting biologist David B. Marshall, reviewed the status of the marbled murrelet. His 1988 report raised warnings that the bird could be in trouble from California to the Canadian border. (He did not look closely at British Columbia, where nesting habitat is falling rapidly to the chain saw, or at Alaska, where there is a large murrelet population.) Marshall found that the Washington population, estimated at between 1,900 to 3,500 breeding pairs, "could be reaching a critical status." The modest bird population faces pressures from two directions: one from continued habitat loss, the other from mortality in salmon fishermen's gill nets. Oregon's somewhat smaller population was predicted to decline with rapid logging of federal lands in the Coast Range. California's population, the smallest of the three states, "is already broken into two segments, each of which are likely to become so small that they could be eliminated from other factors such as oil pollution and gill-net fishing." Neither the Forest Service nor the Bureau of Land Management planned any broad study of how their timber sales might be affecting the murrelet's prospects for survival. The Washington Department of Wildlife had no funding to expand its survey of nesting areas during the 1989 breeding season. Marshall's grim assessment of the situation was that the evidence pointed "to a strong possibility of extinction within major portions of the bird's range if conservation measures are not undertaken."

Marshall, who for thirty years worked for the U.S. Fish and Wildlife Service, had put together the fullest report yet done on this bird of the sea and forest. His warnings were not the first, however. The Pacific Seabird Group, an organization of seabird biologists, had been asking federal agencies for five years to take a close look at the marbled murrelet's habitat needs. Audubon Society chapters in Washington, Oregon, and California, backed up by the Marshall report, petitioned the Fish and Wildlife Service early in 1988 to list the murrelet as a threatened or endangered species. If the agency did list the bird under the Endangered Species Act, it would be required to adopt a recovery plan that could affect federal timber sales. The Audubon Society petition was filed a month after the Fish and Wildlife Service denied a separate petition by Audubon and other organizations to list the northern spotted owl. The spotted owl's problems were far better documented than the marbled murrelet's; there seemed little likelihood that Fish and Wildlife would agree to list the murrelet. The Audubon Society's murrelet petition was calculated to accomplish two things even short of outright victory. First, it would begin to focus public attention on the fact that species other than the spotted owl may be harmed by the federal government's old-growth timber sales. Second, it would make it politically difficult for Fish and Wildlife, the Forest Service, and BLM to continue to ignore the issue of the marbled murrelet; the petition could help spring loose the funds needed to find out whether the bird is in deep trouble.

Whatever action Fish and Wildlife took, the marbled murrelet was on its way to joining the spotted owl as another rallying point for activists seeking to save old growth. There remained the possibility that the issue could be defused. The National Council of the Paper Industry for Air and Stream Improvement (NCASI) began to work with government researchers to find the best way of capturing murrelets and outfitting them with radio transmitters. Birds with radios could be tracked back to their nests. If research showed murrelets thrive in second growth, it would put to rest the idea that mature and old-growth forests must be protected as nesting habitat.

The evidence continued to grow during the 1980s that old-growth forests provide an important, if not absolutely necessary, habitat for a number of vertebrate species. Yet definitive answers remained elusive. How were national forest managers to maintain "viable" and "well distributed" vertebrate populations, as required by law, if scientists couldn't tell them how to do that? The growing political struggle over old growth often turned on scientific issues, but scientific certainty was hard to come by. Researchers couldn't keep pace with the demands of politicians and bureaucrats for hard answers to tough questions. What is the minimum viable population (MVP) for species such as the spotted owl and the marbled murrelet? How much old growth or mature forest is required to maintain that MVP? How should protected areas be laid out on a map?

Scientists accustomed to being ignored by the public suddenly found themselves in the glare of media attention. Congress threw more and more money in the direction of biologists doing field inventories--at least when it came to the spotted owl. Political zealots on both sides eagerly offered the answers that the scientists didn't yet have. Some environmentalists shrilly protested that dozens of birds and mammals would become extinct if the log-ging of old growth continued. Every time a spotted owl was seen roosting in second growth, timber-industry executives pooh-poohed the well-documented link between owls and old growth. Neither of those extreme positions was supported by the evidence. In the absence of scientific certainty, the Forest Service and BLM continued to sell old-growth timber at a pace that alarmed biologists.

There was a time when foresters generally considered coniferous forests--including old growth--to be a "biological desert." Rutherford Platt fell into this trap in his generally excellent tribute to the forests of this continent, *The Great American Forest*. Platt acknowledged that a few grizzlies still ranged the coastal conifer forest, that a cougar might stake out a deer trail, that a marten would chase a squirrel through the treetops only to be hunted down in turn by the incredibly swift fisher. Not bad for a "biological desert." Yet Platt, partial toward the deciduous forests of the East, mistakenly believed all these mammals turned to the forest "only for stalking and cover."

More compelling for her first-hand knowledge is the complaint of a pioneer in the Olympic rain forest about its eerie silence. "It is seven months," she wrote in her journal, "since I recall a songbird." Other pioneers were more unsettled by the nighttime screaming of mountain lions. While there were undoubtedly some dark stands of virgin forest where birds rarely sang, that was the exception rather than the rule. In my own rambles in the old growth of the Cascades and Olympics, I have almost always been serenaded by songbirds. This forested region is home to twice as many breeding bird species as Florida's forests, and nearly three times as many kinds of mammals.

John Muir understood how easily one could overlook the forest animals, most of them small, some living their entire lives reclusively in the treetops. Sitting quietly among the virgin forests of western Oregon, Muir was delighted by the songs of ouzels, thrushes, linnets, and warblers--and by the beating of hummingbirds.

> But few of these will show themselves or sing their songs to those who are ever in haste and getting lost, going in gangs formidable in color and accoutrements, laughing, hallooing, breaking limbs off the trees as they pass, awkwardly struggling through brierly thickets, entangled like blue-bottles in spider-webs, and stopping from time to time to fire off their guns and pistols for the sake of the echoes, thus frightening all the life about them for miles. It is this class of hunters and travelers who report that there are "no birds in the woods or game animals of any kind larger than mosquitoes."

As a native of the East transplanted to the Northwest in my twenties, I understand the prejudice that some people feel about conifer forests. When my father took me hiking and camping in the state and national parks of Virginia, I developed a decided preference for deciduous forests. The hyperactive chlorophyll of their translucent broad leaves produced the most brilliant greens imaginable. By comparison, the conifer forests of my youth were something to be gotten through as quickly as possible on the way to the next meadow or grove of deciduous trees. I didn't know anything about "biological deserts," but I could sense where there was a profusion of life and where there wasn't. The second-growth conifers we encountered in those days were pole-sized pines, all the same young age, growing closely together. The canopy was so solid that the sun's rays couldn't reach the brown needle-covered ground. In their desperate competition for sunlight, the pines put out foliage only at the tops of the trees. Their identical trunks were covered with spindly brown branches that bore no needles. It was a grim, forbidding forest.

That, for most of my life, was my idea of a coniferous forest. Compared with either the riotous colors and shapes of the deciduous forest or the luxuriant moss and ferns of the old growth, those pine forests seem threadbare indeed. I can't help but wonder if some of those who have called the coniferous old growth a biological desert somehow confused some of the world's most magnificent forests with young even-aged stands like the pine woods I disliked so much. The tree farms that now cover most of the Douglas fir region bear far more resemblance to young pine woods than to old growth.

The luxuriance of the old-growth world is well displayed in the Boulder River forest northeast of Everett, Washington. This, the lowest-elevation forest protected by the one-million-acre Washington Wilderness Act of 1984, serves as a reminder of what is unique about the virgin woodlands of the Pacific Northwest.

I'm stepping onto a large log that looks as solid as any in these woods, when *crrrrrrunch*. The log has rotted to the point that it's as soft as its brilliant green vegetative cover. Surface appearances, it turns out, are little guide to what lies under a heavy growth of moss, ferns, and herbs.

Farther up the gentle slope, the ground grows even softer. I sink deep into a mat of moss. Loose, hidden layers of rotting wood, conifer needles, and other duff underlie this field of moss topped with inch-long triangular leaf formations. There are no signs that other animals have trampled this new spring growth. Feeling a bit like a bull in a china shop, I clamber awkwardly onto a large log.

This still-solid log, the remains of a tree that grew for hundreds of years before succumbing, is an ideal perch from which to observe this old-growth cedar grove. I have wandered off the trail, drawn in here by something indefinable: perhaps the large trees, the openness of this grove, the allure of nearly flat land in this steep, narrow valley, perhaps the notable stillness on this breezy May afternoon.

An easy stone's throw away from my perch is another moss- and fern-covered log. That smaller log has been on the ground longer than the giant on which I'm standing. Only two feet in diameter, it provided the seedbed for a hemlock tree that has grown to one foot in diameter. Its back not yet broken by the passage of years, the log forms a bridge over a depression in the ground. Most of the hemlocks that begin life in the rotting wood of a "nurse log" like this one die before reaching maturity. But this tree has been successful. Its vigorous roots don't just snake around the log; they have dropped several feet through the air and found mineral soil. In time, the nurse log will rot away, leaving the hemlock suspended on stiltlike roots that will remember forever the log that shaped them.

Nurse logs are not limited to old-growth forests, but that is where they reach their highest form. They are the preferred seedbed for western hemlock and Sitka spruce. It is common to see lines of young trees atop a log, even colonnades of mature trees following the line of their long-decayed nurse log. This phenomenon is not a new discovery. Explorer Meriwether Lewis remarked on it when he visited the Lower Columbia River: "The hills up this river and towards the bay are not high, but thickly covered with large pine of several species: in many places pine trees, three or four feet in thickness, are seen growing on the bodies of large trees, which though fallen and covered with moss, were in part sound." (Lewis's "pines" very likely were hemlock or spruce.)

The riot of vegetation in this cedar grove typifies the Northwest rain forests. Annual rainfall here on the windward slope of the Cascades may be anywhere from sixty to 100 inches. Tree trunks are completely clad in moss up to about the twenty-foot level. It is as though greenish fairy dust had been sprinkled from the ground up. In addition to the mosses, liverworts, ferns, and moisture-dependent varieties of lichen, fungi of varied colors flourish in these woods. Shelf fungi, dark brown on top and white below, stare out from the trunks of trees like the faceless heads of long-dead Indian hunters.

The log I'm standing on was put here when a mighty tree crashed down the hillside, smashing into pieces. Each piece zigzags at its own angle, the top of the fallen tree leaning against the base of a large cedar. The cedar is king of this grove. Eight or nine feet in diameter at breast height, it is the largest of the cedar. Visually, it is the focus of a natural amphitheater formed by the curving hillside. Douglas fir, which is the dominant tree in other parts of this valley, is absent from this grove of cedars.

Under the cover of the giant cedars, hemlocks and a few younger cedars are struggling to become the dominant trees of the forest. But until the cedars give way to wind or decay, the young trees will remain also-rans. As the big cedars fall, more sunlight streams in and enables the young trees to grow faster. From time to time--though not often in valleys as moist as this--fire sweeps through, killing almost everything in its path. If that should happen to this forest, these cedars and hemlocks could give way to a new stand of Douglas fir.

In the past, when scientists talked about forest succession, everything seemed neat and tidy. The process began with the winds or fires of nature that cleared the old forest and prepared the site for the new. Within a few weeks of the catastrophe, grass and shrubs would begin to turn the land green once again. Trees would emerge next, often red alders and Douglas fir. Growing to pole size, the trees' crowns would form a closed canopy that allowed few shrubs or younger trees to grow in its dense shade. Conifers would replace the short-lived deciduous trees.

A century after the great fire or windstorm, a scattering of trees had fallen, letting some light through the forest canopy. Shade tolerant trees, particularly western hemlock, grew in the few rays of sunlight reaching the forest floor. The forest had reached maturity. During the second century, the dominant trees would grow to considerable size--three, even four feet in diameter. Large snags and logs could be seen here and there. The understory was filled with trees of varying size. The forest had reached the early stages of old growth.

After another 500 to 700 years, more of the pioneering Douglas fir fell. The mature understory hemlocks and cedars became the dominant trees. Able to grow in their own shade, these species would remain dominant. This was a climax forest. Its basic structure would remain unchanged until the cleansing hand of nature brought the next natural catastrophe.

As long as researchers tested the model by looking at what happened after a site was clearcut or burned to a crisp, the model worked quite well. Succession *can* take place in much the way the dominant model describes. Reality is not as simple as the model, however. Nature doesn't often destroy the complexity of old growth and replace it with the simplicity of bare land and tree-farm-type regrowth.

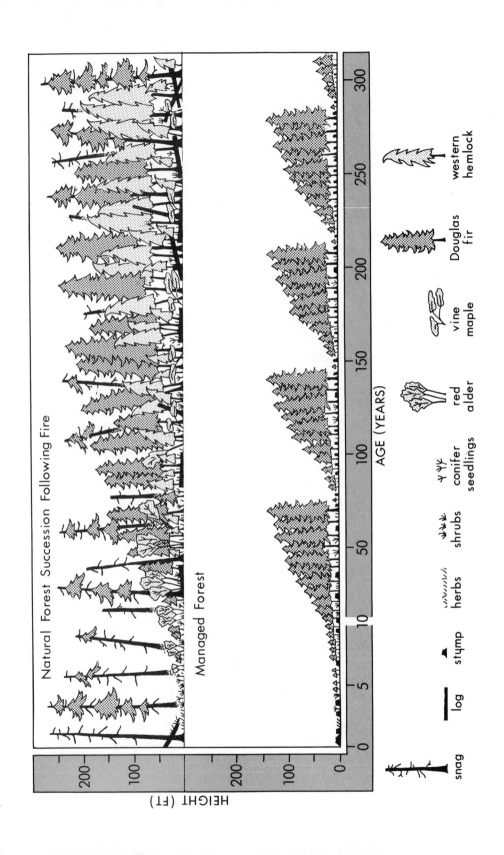

Natural Forest Succession Following Fire

Managed Forest

HEIGHT (FT)

AGE (YEARS)

snag   log   stump   herbs   shrubs   conifer   red   vine   Douglas   western
                                            seedlings alder maple   fir   hemlock

Invariably, a significant amount of material is brought from the old forest into the new. It is only the largest, hottest fires that produce a wholly new forest. More often, on reasonably moist Douglas fir, Sitka spruce, and cedar sites, fire clears out the understory without doing serious damage to the largest trees. The height of those trees protects their crowns, and the fire-resistant bark of fir protects its thick boles. Even the kind of devastating fire that races through the crowns of trees is unlikely to burn all parts of the forest. Some stands are untouched, others suffer partial damage.

In a study of the fire history of two forested areas in the Cascades of western Oregon, two Forest Service researchers found that the frequency of fires varied widely within each study area. Based on the evidence in the growth rings of tree stumps up to 800 years old, Peter H. Morrison and Frederick J. Swanson theorized that three different patterns of fire had shaped the ancient forests. The first scenario was the classic "large stand-replacement fire" that sweeps through a forest every 200 to 400 years. The second pattern was one of catastrophic fires followed by several reburns and then a long period without fire. The third, and most interesting, was the phenomenon of patchy, lower-intensity fires reshaping the forest every twenty to 100 years in some stands.

> Live old-growth trees, large snags, large logs, and high structural complexity may have been sustained in many sites for nearly 1,000 years under the natural fire regime. This is radically different from the classic concept of cycles of complete stand destruction and a sequence of stand development in which old-growth stands occupy a site only a small percentage of the time.

The old forest isn't replaced, it's modified. And the changes are patchy, adding to the forest's complexity. Even when big fires swept across huge swaths of the Northwest landscape, Jerry Franklin observes, "It wasn't like the region, hundreds of thousands of acres, were converted to clearcuts. Rather, you had a lot of heterogeneity as a result of those catastrophes. You can go down and look at the fires that occurred last year in southwestern Oregon. Even where they were very intense, they would leave a tremendous legacy of standing dead and dying material. And almost always you get patches of individual green trees and skipped areas. And very likely it is that complexity of those large disturbances, the heterogeneous pattern, that would have provided a good opportunity for organisms to perpetuate themselves."

Only recently has it become widely appreciated that the biological diversity of ancient forests is giving way to ecological poverty in the young, even-aged forests that now dominate the landscape.

Wildlife management traditionally has been synonymous with game management. Only a few species were of much concern to government agencies: those that were popular among hunters and trappers. The very names of the state wildlife agencies--the Fish and Game Department in California and the Game Department in Washington--spoke volumes on the subject. State game departments were funded primarily by revenues from the sale of hunting licenses, and department managers perceived hunters to be their constituency. When it came to forest management, the species that really mattered were deer and elk. Because it appeared that clearcutting helped those species, many game officials gave their blessing to the destruction of the old-growth forest. Slowly, the idea has grown that nongame species also are important. In 1987, the Washington Legislature dropped the name Game

---

VERTEBRATES ASSOCIATED WITH MATURE AND OLD-GROWTH FORESTS OF THE PACIFIC NORTHWEST

Birds

Goshawk
Northern spotted owl
Vaux's swift
Pileated woodpecker
Hammond's flycatcher
Pine grosbeak
Townsend's warbler
Marbled murrelet
Bald eagle
Osprey

Canopy mammals

Silver-haired bat
Long-eared myotis
Long-legged myotis
Hoary bat
Red tree vole
Northern flying squirrel
Douglas squirrel (chickaree)

Ground mammals

California red-backed vole
Coast mole
Marten
Fisher
Trowbridge's shrew
Pacific water shrew
Shrew-mole
Dusky shrew

Amphibians

Olympic salamander
Tailed frog
Clouded salamander
Ensatina
Oregon slender salamander
Western redbacked salamander
Roughskinned newt
Northwestern salamander

Sources:

Jerry Franklin et al., Ecological Characteristics of Old-Growth Douglas-Fir Forests. *U.S. Forest Service, Pacific Northwest Forest and Range Experiment Station, Gen. Tech. Report PNW-118, February 1981.*

Larry D. Harris, The Fragmented Forest. *Chicago: The University of Chicago Press, 1984.*

Old-Growth Douglas-fir Forests: Wildlife Communities and Habitat Relationships. *Presentations at a symposium sponsored by U.S. Forest Service, Pacific Northwest Research Station, March 29-31, 1989, Portland, Oregon.*

Department in favor of Wildlife Department, acknowledging that there is more to wildlife management than simply catering to hunters. A rare salamander, an owl, or a wolf is just as important to many citizens as a big game animal. As it turns out, what's good for the salamander is good for the elk. The low-elevation, streamside old growth that provides summer moisture for the salamander also offers the best winter foraging for elk and deer.

Old thinking does not die easily. The notion persists, particularly among loggers, that clearcutting boosts deer and elk populations. One acquaintance of mine insists that deer are more abundant today than they were when white settlement began. There is at least a kernel of truth to his belief. A century ago, when old-growth forests were the dominant feature of the Pacific Northwest landscape, logging probably was helpful to deer and elk. Acre for acre, the vegetation in clearcuts offers more food for ungulates than does forest foliage. In the warmer coastal areas, the forage in meadows is available year-round. Not so in much of the Cascades and Olympics and the coastal mountains of British Columbia and Alaska. There, where winter snows are heavy, forests play a critical role in sheltering and feeding large mammals. The herbs, shrubs, and Douglas fir seedlings in a clearcut don't do herbivores much good when the plants are buried under five feet of snow.

The abundance of deer in North America in the colonial era is amply documented. Thirty thousand deerskins were shipped to England from Savannah, Georgia, in a single year, and 132,000 skins from Quebec in one year. A report published by the state of West Virginia in the 1950s noted that deer were reported to be "abundant in virgin forests. . . . Virtual disappearance of the deer from most of the State coincided with the big timber-cutting years between 1880 and 1910."

In the forests of the Pacific Northwest, where ancient forests still exist, scientists are learning much about the habitat needs of large game animals such as deer, elk, and mountain goat. For example, snowfall is a critical factor in determining the size of the Sitka black-tailed deer population in southeastern Alaska. Up to 50,000 deer may starve or freeze to death during an especially severe winter. In one study, deer on Alaska's Annette Island were found to use old-growth stands seven times more frequently than clearcuts or second growth in winter. The more continuous canopy of second growth may keep out snow as well as old growth, but it offers a relatively bare cupboard. Shrubs preferred by deer are relatively rare in second growth and few lichens grow on the young trees. In a mild winter, deer will forage in a variety of habitats. But during deep snowfalls, they seek the refuge of older forests. For one thing, the old-growth microclimate provides better "thermal cover," or warmth, than a younger forest. (One reason for the spotted owl's preference for ancient forests may be the protection offered by the forest canopy in severe weather. In warm or hot weather, the owl tends to roost in the cool understory in hemlock or small hardwood trees. In winter, the bird perches in large old-growth conifers.)

The food supply offered by old growth may be more important than thermal cover to deer. Even when snow covers leafy plants on the forest floor, edible lichens fall from the canopy. Beard lichens accounted for more than one-quarter of the diet of a group of deer wintering in a mature forest on Vancouver Island. Four-fifths of the edible material falling to the forest floor consisted of lichens.

No one is suggesting that black-tailed deer, Roosevelt elk, and mountain goats will survive as species only if old growth is preserved. The survival and, in many cases flourishing, of deer on the East Coast demonstrates their ability to survive in younger forests and in clearings. The surprise, to those of us indoctrinated in the old saw that deer prefer clearcuts, is that they rely on older forests at all. If deer, of all animals, benefit from ancient forests, how many other species are even more dependent on them?

The answer to that question is slowly emerging. While Jerry Franklin and his colleagues at the Forest Service research station in Corvallis, Oregon, were beginning to describe and define the structure of old growth in the 1970s, something else was stirring on the Oregon State University campus just a two-minute walk from the Forest Service lab. The late Howard M. Wight, leader of the U.S. Fish and Wildlife Service's cooperative research unit, was grappling with the research findings of two of his graduate students. Richard Reynolds's research showed that the goshawk, a short-winged predator, had a strong preference for older forests. Eric Forsman, tracking spotted owls, discovered they were primarily old-growth creatures.

Researchers began to look more broadly at ancient forests as habitat for a range of species. By 1974, Wight had put together a matrix showing the habitat preferences of ninety-one bird species in ponderosa pine forests east of the crest of the Oregon Cascades. He displayed each bird's use of five "seral stages" in forest succession: grass and forbs (1 to 7 years after fire or clear-cutting), shrub and sapling (8 to 15 years), second growth (16 to 60 years), "older second growth" (61 to 199 years), and "mature" forest (200 years or older). Only "significant use" of a habitat type--not a single stray sighting--was recorded. Some species, such as the song sparrow and the rufous-sided towhee, nested and foraged only in the grass-forbs and shrub-sapling stages. Others, like the black-headed grosbeak, used everything from the shrub-sapling stage on up to old growth. Still others, like the mountain bluebird, were found in all seral stages but nested only in older forests. Each species had its distinctive niche, which generally included several seral stages.

A clear pattern emerged from the work of Wight and other research-ers. Species diversity was greatest in the young and the old seral stages, least abundant in young second growth. Twice as many species were found in "older second growth" and in "mature" forests as in young second growth. And although a wide range of species put in regular appearances in the shrub-sapling stage, that stage was no better for nesting than young second growth. When it came to nesting, older forests were by far the richest habitat. More than one-third of the birds nesting in late-succession forests laid their eggs in the cavities of the snags so distinctive of old growth. Seven woodpecker species, two hawks and two eagles were among the species that nested only in the two oldest forest types. Wight's systematic look at bird habitat echoed Jerry Franklin's observation that old growth is "not just a younger forest grown up to a larger size."

The implications for forest management were immediately apparent. Logging of the ancient forests eliminates the best nesting habitat. Short-rotation tree farms cover most of the landscape with young even-aged forests that provide the biologically least productive habitat. The birds whose futures were "most precarious," Wight observed, were those that, like the spotted owl, were narrowly adapted to older forest. "I consider a stand mature when it contains some trees that are 200 years old or older," he wrote:

> I have heard some timber managers call such stands "decadent." It may be decadent from the viewpoint of the size of the annual increment of merchantable timber, but to a few species of wild-life, it may be their last hope for survival in a natural setting. A good reason for this is that a mature stand of timber is a very stable environment. Over the eons, certain species have evolved in har-mony with this stable condition, and in the process of evolution, little need has arisen for these species to be highly adaptable to rapid changes in the habitat.

More recent research confirms what Wight found. Early in 1989, scientists reported the results of a massive Forest Service-sponsored research effort. Resident birds found their best winter habitat in old growth. More than half a dozen bats strongly preferred old growth. Research suggested that survival of the red tree vole and Vaux's swift are dependent on old growth. The vole, which eats Douglas fir needles, needs large Douglas fir while the swift nests in large, hollow snags.

Had the scientists compared old growth with younger managed forests rather than with young natural forests, the habitat value of old growth would have been even more apparent. Large dead trees, conspicuously absent in most managed forests, are a hallmark of natural forests. After a Douglas fir dies, it goes through stages of decay, each of value to a different set of animals. In the first years after death, wood-boring beetles invade the bark and outer layers of wood. Fungi and bacteria brought in by the insects begin or hasten the process of wood decay. Woodpeckers feed on beetles.

After six or seven years, typically, beetles begin to bore deep into the heartwood. Wood rots where they have dug, and large pieces of the weakened snag crash to the ground. Meanwhile, woodpeckers and nut-hatches excavate nesting cavities in the softened wood. The tree's broken top exposes the heartwood to rapid rotting. The natural cavities of this "soft snag" provide homes for more animals, such as the small brown creeper, Vaux's swift, raccoon, and black bear. One virtue of a snag is that its thick walls moderate the temperature, offering animals a warmer environment in the cold winter months and a cooler environment in the summer. For some species, such as pileated woodpecker, it is important that the nest tree be of large diameter.

A fallen tree provides wildlife habitat even longer than a snag. A thirty-inch-diameter Douglas fir log may take 500 years to become 90 percent decayed. Like a snag, the log is colonized initially by wood-boring insects, fungi, and bacteria. Different varieties of fungi become dominant as the log slowly rots. Its surface is covered with lichens, mosses, and liverworts. Ferns, huckleberry, hemlock, and spruce put their roots into cracks in the increasingly damp wood. Carpenter ants and termites proliferate along with beetles. Earthworms are found as far as six feet above the ground, eating their way through hemlock and silver fir logs. Whether in a log or in the soil, worms break down organic matter, thus increasing decay activity by fungi and bacteria.

During the later stages of decay, the heartwood of a rotting log cracks, making way for even larger populations of fungi and insects and the roots of "nursing" hemlocks. Centipedes, pseudoscorpions, and spiders move in to feed on the smaller mites and ants. The wood turns a deep reddish-brown and holds increasing amounts of moisture. I have grabbed a handful of rotting wood from one of these logs and wrung drops of water from it. The sponge-like ability of a decaying log to hold moisture is critical to the forest ecosystem during the summer when there is little rainfall in the Northwest. The Oregon slender salamander turns to the coolness and moisture of rotting logs to survive during the dryest part of the summer. The Oregon salamander often lays its eggs under the bark or in the decaying wood of Douglas fir logs. The California red-backed vole is fond of hiding under logs and feeding on the truffles that are so abundant there.

As the log crumbles, it may be held together for years by the roots of trees that began life in the log. Eventually, it succumbs and becomes part of the forest soil. After decades of work by nitrogen-fixing bacteria, the log's remains have become a rich fertilizer. From death comes life. The ecological richness of old growth is built on the very "decadence" that some observers deride.

With the possible exception of the dusky-footed woodrat, there is probably no mammal for which young forests are the preferred habitat. Most animals find their optimal habitat either in the very earliest or the very latest successional stages. In the Douglas fir region, researchers Larry D. Harris, Chris Maser, and Arthur McKee identified thirty-six mammals, amphibians, and reptiles whose "primary habitat" is in older forests. Among the mammals preferring old growth were marten and fisher, both members of the weasel family; several varieties of bats; two voles; two squirrels; bushy-tailed woodrat; and coast mole.

Wildlife populations in old-growth forests reflect those forests' distinctive food chains. The red-backed vole is the only species in North America known to feed almost exclusively on fungi. The northern flying squirrel in the Douglas fir region is the only North American mammal that feeds principally on lichens. (In other parts of North America, where the flying squirrel also lives, it eats large seeds.) Snakes and salamanders abound in part because of the bounty of insects that feed on decaying wood.

What these woods lack, like many other forests, is an abundance of herbivores. The old-growth forest's unusual food chain favors fungivores and carnivores. Roughly two-thirds of the birds found in old growth consume insects or higher animals. Sixty-five percent of the mammals, amphibians, and reptiles in the Douglas fir region are carnivorous. Of the twenty-two mammal species weighing more than one kilogram, Larry D. Harris and Chris Maser report that twelve are true carnivores and only five are herbivores. In the Rockies, by contrast, the ten largest species include one carnivore, one omnivore, and eight herbivores.

The preference of most wildlife for the earliest and latest seral stages suggests that the full impact of logging isn't felt for several decades. When old growth is cut, one wildlife-rich landscape is replaced with another. The brush and seedlings that dominate a site after clearcutting support many vertebrate species. (However, many of the animals that thrive in older forests are absent in this altered environment.) Wildlife diversity plummets within a few years as a young forest emerges with its unbroken, unvarying canopy. In "managed forests," trees are perpetually of a uniform age and, most often, a uniform species. This kind of plantation lacks the structural diversity that supports a profusion of life in older, natural forests. As zoologist A. Starker Leopold reputedly put it, an even-aged stand of Douglas fir offers such sparse food sources that "a bluejay would have to pack a lunch to get across."

Harris and his colleagues reported in 1982 that only one-fifth of the publicly owned forestlands of western Oregon and western Washington were in the most biologically barren seral stages. But that was rapidly changing:

> Fifty years from now as much as 65 percent of public land acreage
> and a higher percentage of total acreage will fall between the ages
> of 30 and 90. Unless remedial steps are taken, wildlife population
> declines, similar to the declines in midwestern and eastern areas
> earlier this century will probably occur.

The total amount of forest land in various age classes tells only part of the story. Biological diversity is greatest at the lower elevations. Forty percent more species of amphibians, reptiles and mammals are found at 1,000 feet above sea level than at 4,000 feet, according to Harris, Maser, and McKee. Before logging days, the most extensive, biologically productive forests in the Douglas fir region probably were those of northwestern Oregon and southwestern Washington. These are almost entirely in private hands, and were among the first to be logged over. Virtually nothing remains of those legendary woods. In the national forests, much of the original forest cover remains, but logging has been heaviest in the valley bottoms where the first roads were built. Congress has designated wilderness and established parks mostly at the higher elevations, where species diversity is poorest.

Because deer and songbirds may be more plentiful in a clearing than in the deep forest, some people have leaped to the conclusion that clearcutting makes for better wildlife habitat. Total animal populations and diversity of species are two different things. The richness of an ecosystem turns not so much on how many animals are present as on its diversity of species and their relationships with one another. Yet for most of this century, foresters and game managers have used the populations of nonthreatened species like deer as their yardstick for measuring biological diversity. The conclusions drawn from that standard have made as much sense as regarding the flocks of pigeons on downtown streets and of starlings in the suburbs as evidence that bird habitat is improved by paving over wetlands.

The most comprehensive investigation into vertebrate use of the old-growth ecosystem is a Forest Service project. Project leader Len Ruggiero has the sort of don't-mess-with-me air one might expect of Mike Tyson. Perhaps he's always had that look. Or perhaps it's developed over the years he's tried to direct forest ecology research in an agency that spends its biggest bucks administering timber sales. In bureauratic terms, Ruggiero's project is the Pacific Northwest Research Station Research Work Unit (RWU) 4203, Wildlife Habitat Relationships in Western Washington and Oregon. Better known as the Old Growth Project, RWU 4203 is housed in a small Forest Service research complex in a second-growth forest near Olympia, Washington.

Ruggiero's sky-blue eyes bore into me as he tells how the Old Growth Project has been blown about in shifting political winds since its creation in 1982. The research and development program was set up for the purpose of exploring the use of mature and old-growth forests by a variety of vertebrates. The spotted owl would be its top priority, but the project would study other birds, mammals, amphibians, and reptiles as well. When the lumber market collapsed in the early 1980s, so did much of the project's funding.

Meanwhile, the National Audubon Society and other environmental groups were lobbying Congress to fund a larger research effort on the spotted owl's habitat. Congress obliged with a $1.5 million appropriation. Rather than channel the new spotted owl funds funds through Ruggiero's old-growth habitat project, the Forest Service created a new research and development program aimed exclusively at the owl. RWU 4203 receives considerable support to study the spotted owl (about 70 percent of its budget), but there aren't enough dollars to study the dozens of other species that show an affinity for older forests.

"This is somewhat short-sighted," Ruggiero argues. "I think there are other issues that are important and that are going to become more important in the future--perhaps other spotted owls, perhaps other ecosystem-level concerns--and we're really not paying the type of attention we should be. There are only so many resources and they're being channeled to the spotted owl."

It hasn't always been helpful to the careers of government scientists to point out the damage that logging does to a range of wildlife. Chris Maser, a biologist for the Bureau of Land Management, was called across the country to BLM headquarters to explain his public statements that raised questions about the agency's timber sales. Maser, an independent soul, chose to view the inquisitors in the director's office as a "captive audience. . . . I was told, 'You have to understand the effective policy.' My response was that that was not my concern. My responsibility was to report the truth." Maser later left BLM.

Even in the Fish and Wildlife Service, some scientists are careful about what they say. Fish and Wildlife is part of the Interior Department, the parent department of BLM. "I'll get my butt burned--bad," I was told during the Reagan administration's second term by one government scientist afraid to speak too candidly. Managers of federal agencies, concerned about the political ramifications of their staff scientists' research, sometimes refused to authorize expenses for travel to meet with colleagues in other agencies. "Agencies can make it very difficult for people to participate in the process," said the same scientist, who asked to remain anonymous.

Even with generous funding and more supportive federal agencies, it wouldn't be easy to define the extent of many animals' dependence on ancient forests. How, for instance, are scientists to track the movements of a tiny Vaux's swift as it darts through the old-growth canopy at more than forty miles per hour? In some ways, the marbled murrelet is even more elusive. It took a half a century and a lucky break before ornithologists were able to prove the bird nests in big trees. Researchers from the Old Growth Project didn't encounter any murrelets in the first year of their population surveys; had they not known better, they could easily have concluded that these are not forest birds.

The goshawk is another difficult bird to study. This bird seems to prefer late-succession forests for nesting, while its cousin the sharp-shinned hawk is found in early-succession forests and Cooper's hawk in intermediate stages. Further research is needed to confirm that apparent pattern.

Rich Lowell, a biological technician, is gathering data for Ruggiero in the Olympic National Forest's Quilcene Ranger District to determine if goshawks can be effectively tracked by radio telemetry in this rugged landscape. Only two and a half weeks before my visit, he and the three men working with him attached radios to an adult male and one of its offspring, a fledged juvenile male. Already, they can see the job won't be easy. "The adult doesn't sit still very long," Lowell explains as we bump along a dirt road on the northeastern corner of the Olympic National Forest. "He's jumping around all the time."

Over the next five hours, Lowell and Jim Swingle work to pinpoint the location of the adult male. Lowell is on one side of the valley, Swingle on the other. With their directional antennae, the two researchers simultaneously check the bird's bearings. Then Swingle moves to another location and takes another reading. Drawing lines on their maps, they try to pinpoint the bird's location. On their first several tries, the bird is considerably to the north of both men. When they draw three lines on their maps, the lines don't intersect to give a precise fix on the bird. The steep terrain sometimes confuses the radio signals. Then the hawk moves. Swingle is closer to it, so Lowell turns around on this dead-end logging road, descends into the valley, and back-tracks northward to get in a better position. After hours of repeated move-ment by the two men and by the hawk, the researchers are satisfied that they've gotten a few reliable readings. In the process, they've learned a few more lessons about working together and about this valley's terrain and road system.

Over the past several months of stalking goshawks and now following them with the help of radios, Lowell has been impressed with how "secretive" the birds are. "I think it fits into their hunting strategy to be an ambush kind of hunter," he says. "Some birds, like the red-tail, soar above the canopy. They're highly visible. The goshawk is really down in the timber below the canopy. We rarely get a good look at a goshawk, even those of us who are out here studying them."

More difficult than doing the research on the old-growth habitat is the task of translating the fragmentary research into management decisions. The National Forest Management Act (NFMA) of 1976 directed the Forest Service to "provide for diversity of plant and animal communities." It was left to the Agriculture Department, parent agency of the Forest Service, to write specific regulations to implement this and other NFMA goals. The regulations were adopted six years later, in the fall of 1982.

The NFMA regulations, based on the recommendations of a committee of scientists, put the meat on the bones of the diversity mandate. "Viable populations" of all native vertebrates are to be maintained. The regulations recognize that habitat is the critical factor: "In order to insure that viable populations will be maintained, habitat must be provided to support, at least, a minimum number of reproductive individuals and that habitat must be well distributed so that those individuals can interact with others in the planning area." The adequacy of habitat is to be determined by selecting and monitoring the populations of "management indicator species." Indicator species may be chosen because they are threatened or endangered, because they have special habitat needs, because they are commonly hunted or fished, because they are species of "special interest," or because they will indicate the effects of logging and other activities on wildlife.

Three species have been chosen as indicators linked to mature and old-growth forest habitat west of the Cascade crest: spotted owl, pileated woodpecker, and marten. In some forests, deer, elk, and mountain goat are indicators for the forest cover that many species require in winter. The bald eagle, because it is on the government's endangered species list, also is an indicator.

The three old-growth indicators were chosen because of their association with late-succession forests and because of their large home ranges. If enough habitat is protected to meet their needs, the Forest Service reasons, then the needs of dozens of other species that also use older forests will be met. For each of the three indicator species, a network of habitat areas has been established.

The pileated woodpecker, a grand bird distinguished by its size and its red crest, is the largest woodpecker in North America after the ivory-billed woodpecker. The ivory bill, a somewhat larger bird, serves as an object lesson for what could happen to its red-headed relative. Once spread over much of the southeastern United States, the ivory bill thrived only in virgin hardwood forests. Heavy logging between 1885 and 1915 reduced the bird's range to a few isolated stands. Finally, the last stands were logged. No breeding ivory bills have been seen in this country since 1939. The bird was thought to have gone extinct until a few ivory bills recently were found in Cuba.

The pileated woodpecker seeks large snags in which to excavate a home; when abandoned, these holes may be colonized by saw-whet owls, flammulated owls, kestrels, flying squirrels, and other animals. Its favorite food, carpenter ants, is abundant in the rotting snags and logs of old-growth forests. The pileated woodpecker's range spans the continent. Its eastern habitat decimated by logging, ornithologists at one point feared for its survival. However, the bird has reestablished its population in the East, possibly due to conservation measures or possibly due to adaptation to a changed environment.

The marten is among the world's fastest treetop racers. Valued for its lush fur, this member of the weasel family has suffered at the hands of trappers as well as loggers. The marten was extirpated from much of the Great Lakes region after the heyday of logging there and was nearly wiped out in Ontario. In recent years, it has made a comeback. Like the pileated woodpecker, this skilled predator prefers older forests. Snags and logs provide dens as well as hunting areas. The marten's faster relative, the fisher, is usually found at lower elevations. Trapping and the near-complete logging of its habitat have extirpated the fisher from most, if not all, of its former range in coastal Oregon and Washington.

The bald eagle typically builds its nest near the top of a tall Douglas fir or Sitka spruce in an old-growth or mature forest. Primarily a predator on fish, eagles congregate in the trees along rivers supporting big salmon runs. After fishing on Washington's Skagit River on a cold winter day, the birds will fly several miles to roost in the relative warmth of an old-growth fir forest.

The large amount of old-growth or mature forest used by the northern spotted owl--an average of more than 2,200 acres per pair in central western Oregon and 3,800 acres in Washington--suggests that this species will continue to be the focus of the debate over old-growth habitat for the foreseeable future.

As long as some animals are going to picked as surrogates for others that live in old growth, the Forest Service's choices of the pileated woodpecker, marten, and spotted owl are reasonable. But many scientists are uncomfortable with the whole idea of indicator species. If the objective is preserving old growth as an ecosystem, they argue, then the emphasis ought to be on the ecosystem, not on a handful of its occupants. The current species-by-species approach to habitat management has led to a costly and often frustrating search for "old-growth obligates," or animals that can live only in old growth. Scientists have yet to find any animal whose survival is incontrovertibly tied to old growth. What they have found is an impressive number of species that are most abundant in older forests and that depend on forest characteristics that find their highest and best expression in old growth. For some species, it is hard to determine which old-growth characteristics are most important. Scientists' inability to point confidently to specific animals as old-growth obligates is no reason for complacency. As the landmark 1981 report, *Ecological Characteristics of Old-Growth Douglas-Fir Forests* noted:

> Whether any species depends totally on old growth for survival
> is not clear; however, the fact that a species can survive in other
> age classes of a forest does not necessarily mean it can survive
> once the major reservoir of optimum habitat is gone.

Len Ruggiero warns of a potential "ecological trap" in using the spotted owl as a surrogate for some very different kinds of animals. Protected areas will be closely tailored to the indicator species. "I think it's safe to say you could manage old growth for spotted owls and meet the needs of spotted owls while failing to meet the needs of, for example, some of the aquatic amphibians that use headwater streams in older forests," Ruggiero says.

When the Forest Service and the Bureau of Land Management set up habitat areas for the protection of indicator species, the tendency is to create as few areas of as small a size as it is believed will sustain the species. Both agencies are expected to sell timber, and every acre of protected forest is an acre of timber that can't be logged. It is questionable whether current policies are adequate to maintain viable populations of several indicator species. Environmentalists went to court in 1987 in an attempt to stop four national forests in Texas from logging around the nesting colonies of the red-cockaded woodpecker. This woodpecker, once abundant throughout the southern pine forests, has declined with the disappearance of the large trees in which the bird excavates its nest. In the Northwest, Forest Service biologists give the spotted owl a "low" chance of long-term of surviving in most of its range under current logging plans. The National Council of the

Paper Industry for Air and Stream Improvement (NCASI) says the Forest Service is following a "risky" policy in its plan to protect the marten. The protection plan disperses across the landscape 160-acre patches of old growth--patches far smaller than the marten's measured home range of 576 to 2,048 acres. When an *industry* group says a habitat area is too small, something must be wrong.

On the planning maps of the national forests in the Douglas fir region, commercial forestlands are dotted with colored spots that represent pro-tected habitat for marten, pileated woodpecker, and spotted owl. The discon-nected habitat islands are spread out as widely as possible in a questionable effort to comply with the NFMA requirement that animal populations be "well distributed." The flip side of these habitat islands is that they seem almost calculated to isolate populations; certainly they hasten fragmentation of the ancient forests that are the animals' primary habitat.

Scientists' concerns about ecosystem fragmentation have focused on the Forest Service's policy of logging in patches of forty acres or less in the western Oregon and western Washington. Jerry Franklin teamed up with Richard T.T. Forman of Harvard, the dean of landscape ecology, to analyze the effects of dispersed clearcutting. Noting that "forest edge" is a very different environment than "forest interior," the scientists found that the ecological character of the uncut patches is drastically changed by patchwork cutting. By the time 30 to 50 percent of a drainage has been cut into twenty-five acre patches, Franklin explains, "You don't have any interior forest environment left." The population of forest interior plant and animal species plummets. Alteration of the forest interior might have little effect on deer and elk, which move easily in and out of the forest interior. Spotted owl, marten, red tree vole, and Olympic salamander, by contrast, reap no benefits from edge--and appear to be threatened by it. For them, patchwork cutting could mean disaster.

The calculations by Franklin and Forman assumed that everything would go *well*. An old-growth habitat area or a forested "leave strip" between clearcuts is vulnerable to windthrow. In the 1983 windstorm that ravaged the old-growth forests of the Bull Run watershed east of Portland, Oregon, 81 percent of the 2,200 acres of blowdown were adjacent to roads or clearcuts, the researchers reported.

Those findings led Franklin to a conclusion that bordered on environ-mental heresy: if the Forest Service is going to continue selling old-growth timber, the forest should be cut in larger units, thus leaving larger, more viable stands intact.

National parks and wilderness areas are widely viewed as effective refuges for animals whose original ranges have been depleted by logging, farming, and urban development. But it is becoming increasingly apparent that those reserved areas are inadequate for many species. Their high elevation makes them inhospitable to many species, and there is growing evidence that they can't support carnivores with large home ranges. A grizzly bear doesn't display the owl's finicky need for a special kind of forest, but may range over several hundred thousand acres. Wolf packs travel over huge areas as well. Some of the largest and fiercest carnivores are close to extinction in the Cascades: grizzly, gray wolf, fisher, lynx, and wolverine. The spotted owl is not far behind.

Several researchers have reported the apparent extinction of a number of mammals in Mount Rainier and other national parks of the West. Scouring the records of eighteen parks, William D. Newmark reported in *Nature* that forty-two species apparently had gone extinct. The lost species were those that use the later successional stages such as mature and old-growth forests. Newmark concluded that the parks' ability to support mammals was crippled by "the active elimination of fauna on adjacent lands." Those adjacent lands were primarily national forests. The lay term for "elimination of fauna" is "logging." With massive habitat loss on the adjacent lands, the parks simply weren't large enough to maintain native mammals, according to Newmark.

His conclusion did not go unchallenged. James F. Quinn of the University of California at Davis teamed up with Charles van Riper III of the National Park Service and Hal Salwasser of the Forest Service to gather additional data from biologists and officials in the parks. They came up with results at odds with Newmark's. Some of the "extinct" species, they reported, had never had breeding populations in the parks and there was evidence most of the other species were still present. That left only Mount Rainier's wolves and the river otters of the Sequoia and Kings Canyon parks as possible extinctions. Even those animals, the authors said, might still be present in the parks. In a separate article, Quinn commented on the poor historical records, which threw an element of uncertainty into both the Newmark and the Quinn findings.

Quinn acknowledged an "alarming" extinction rate among North American mammals generally, but felt there was no basis for Newmark's suggestion that an effective response would be to maintain habitat on the periphery of the parks. Quinn and his associates argued that protection might be more effective by continuing to provide smaller habitat areas distributed over a wider landscape.

Behind all the bombast of the conflicting research a common thread ran through the Newmark and the Quinn papers. Both pointed up the critical importance of mammal habitat on the national forests. The Quinn group rejected eleven claims of extinction in the parks because there was evidence that the species in question were present in the adjacent national forests. The authors attributed the animals' absence in the parks to availability of superior habitat in the national forests. That conclusion echoed what other researchers were saying throughout the 1980s: the best forest habitat remaining is what's at the lower elevations in the national forests. And that habitat is disappearing fast.

We don't know just how much natural forest habitat is needed to maintain a viable ecosystem with healthy plant and animal populations. All we know for sure is that warning bells are ringing loudly.

By the end of the 1980s, a consensus was growing among forest researchers that protection of species dependent on older forests must be a two-part affair. One part is preservation of an adequate amount of natural forests in all stages of succession, but especially in old growth. The other part, recognizing that most of the natural forests have been lost, is incorporation of more structural diversity in managed forests. Some scientists, viewing the plight of the spotted owl and some other creatures as truly desperate, argued that all logging should be banned, at least temporarily, on late-succession forests. Others, more sanguine, felt adequate habitat could be maintained through a blend of pristine habitat areas and more environmentally sensitive logging.

One group of forest ecosystem researchers--several of them on the Forest Service payroll--gave some thought to our current effort to protect creatures of ancient forests on a species-by-species basis. Without even understanding an animal's life history, we try to define its minimum needs and then scatter isolated "habitat areas" about a mostly clearcut landscape. The real issue, the scientists pointed out, is not the protection of one little-understood species or another. It's an endangered ecosystem. How much old growth, how much natural forest, is needed to preserve the ecosystem? Here's how Jack Ward Thomas, Len Ruggiero, and three co-authors addressed the question:

> The answer to 'How much?' must be predicated on the relatively small amount of unevenly distributed remaining old growth and the current, inconclusive scientific knowledge of old-growth ecosystems. Therefore, the best probability of success is to preserve all remaining old growth and, if possible, produce more.

Larry Irwin, biologist for NCASI, agreed that scattered habitat areas keyed to specific species was fraught with risks. Even if those habitat areas attracted the desired species, Irwin warned, they could become "reproduction sinks" from which juveniles would disperse but fail to find mates. Far better, he said, to manage larger tracts of forest in more creative ways. Tree farms incorporating elements of the natural forest could offer "minimally suitable habitat" that would do a better job of meeting the animals' needs for habitat and reproduction. Irwin stressed that old-growth-dependent species such as the spotted owl also need an ample reservoir of their primary habitat. The key elements of such minimally suitable habitat are reasonably well understood: leave substantial numbers of large live trees, snags, and logs.

Not everyone cherishes that sort of disorder. Bill Hagenstein, executive vice president of the Industrial Forestry Association, once said, "Forestry has always been an environmental undertaking. Its main thrust has always been taming the ecology instead of letting it run rampant as though there were no people around."

After the catastrophic Yellowstone National Park fires of 1988, some timber-industry lobbyists seized upon public concern over the fires to argue that ecology should not be allowed to run rampant either in parks or wilderness areas. The National Park Service's "let-it-burn" policy sparked particular outrage. Gus Kuehne, president of Northwest Independent Forest Manufacturers, cast the issue this way:

> If environmental extremists are so concerned with protecting the remaining old growth that exists in the State of Washington on lands where it is available for timber harvesting, why are they so anxious to see it burn in areas protected from timber harvesting? Spotted owls, pine martens, marbled murrelets, pileated wood-peckers, and other critters, which biologists believe require old growth, don't know or I submit don't care whether those trees are consumed by wildfires or harvested for a variety of useful products.

In fact, a pileated woodpecker undoubtedly *does* care whether its habitat is altered by a wildfire that leaves behind snags for nest trees or by intensive timber management that will never grow trees large enough for a nest. Kuehne suggests that fire suppression could increase the amount of old growth on the reserved lands. Younger age stands could grow to old growth, and the existing old growth could remain perpetually old. That would provide more old growth habitat, so loggers could cut more wood without environmental damage.

It's an attractive argument. Unfortunately, nature doesn't work that way. Forests, like any biological system, are in a constant state of change. "Old growth is but a stage in forest development," notes Andy Carey, research wildlife biologist for the Forest Service's Old Growth Project. If a grove of giant Douglas fir remains undisturbed by fire for a millennium, its giant Douglas fir will inevitably give way to smaller successors. In all likelihood, the climax forest that replaces the old-growth stand will provide optimal habitat for fewer species.

No one can say with certainty how a forest will change in the future. All we can say for sure is that the forest will change. No matter how we may try through human intervention to "preserve" the forest, it cannot be kept as it is now. It may become a climax forest, it may repeat the process of forest succession beginning with a nearly bare site. Periodic wildfires may even keep some stands in an old-growth condition for a millennium or more.

We might make ourselves feel better if we sent fire battalions into the wilderness and parks in an effort to save the old growth. It's doubtful, however, that we could improve on nature. During the last two decades,

Top: An inhabitant of the big trees, the marbled murrelet, nests in a mountain hemlock in this rare photo taken at Kelp Bay, Alaska. Bottom: These mossy Sitka spruce on the Olympic Peninsula's Bogachiel River aren't really joined together by a root. Rather, the root, part of the tree on the right, ran along a nurse log that decayed long ago.

we've identified some of the features that make old-growth habitat so unique and rich. We've also learned that natural forests almost always produce habitat superior to managed forests. Do we now believe we can do a better job than nature at producing habitat for spotted owls, marbled murrelets, and pileated woodpeckers?

The melancholy cry of the marbled murrelet continues to haunt me. There's mystery and, it seems, a touch of the absurd in the call of a seabird so far from the water. It was by chance, not through any human design, that the presumed nest trees of the South Fork Stillaguamish's marbled murrelets have been spared. What other surprises are hidden from us in the treetops or beneath the rotting logs of the old-growth forests? Will the mysteries still be there when we're clever enough to crack them?

In his essay, "Thinking Like a Mountain," Aldo Leopold imagined that only a mountain has lived long enough to decipher the secret meaning in the howl of a wolf. When Leopold and a companion come upon a pack of wolves playing in a river, they picked up their rifles and shot wildly:

> In those days we never heard of passing up a chance to kill a wolf. ... We reached the old wolf in time to watch a fierce green fire dying in her eyes. I realized then, and have known ever since, that there was something new to me in those eyes--something known only to her and to the mountain. I was young then, and full of trigger-itch; I thought that because fewer wolves meant more deer, that no wolves would mean hunters' paradise. But after seeing the green fire die, I sensed that neither the wolf nor the mountain agreed with such a view.

Years later, after wolves had been killed off in state after state, Leopold saw mountains stripped bare of their vegetation by deer herds unthinned by wolves. He then understood something of the secret meaning in the wolf's howl.

We don't easily pass up the chance to cut a tree. We have plenty of reasons: paying mortgages, buying cedar decking more cheaply, providing better forage for deer. Perhaps the forest hears the secret meaning in the cry of the murrelet. A meaning we haven't yet discovered.

5

The Lord's Crop

The sawyer eyes the grain showing at the end of the large log. It must be turned at just the right angle for cutting five-inch-thick slabs that can be resawn into vertical grain lumber. This log presents a special problem because of a defect in the heartwood. Working around the defect as best he can, the sawyer puts the carriage in motion. First moving forward, then backward, the log passes through the band saw. Slices fall off like butter, slices so heavy most men couldn't lift them.

Suddenly, the earsplitting noise of the saw shrieks to a new high. Then silence. A chunk of defective wood has fallen off the log and caught on the saw. Workers scurry to replace the damaged blade so production can resume.

Big logs are cut up in the "grade lumber" mill at Summit Timber Company much the way they've been sawn in old-growth mills throughout this century. The product is one whose value is as much aesthetic as functional. If this knot-free wood is to look good--and meet grading standards--the grain must be tight. When a two-by-five board is laid on its side, the grain must stand vertically.

There isn't any computer that can figure out the best way to cut grade lumber. It's an art requiring the judgment of an experienced sawyer. But in cutting wood from a natural forest, defective timber is part of the game. "Wrecks" like the one that just occurred are not uncommon. The rewards are high, however. The steeper labor costs and slower production pace are more than offset by the market value of a top-quality product. While the dimension lumber mill next door will bring $300 or less for a thousand board feet of second-growth lumber, the old-growth grade lumber here brings $1,000.

Bill Walter, the quality control supervisor, stands in a light drizzle showing off a stack of defect-free two-by-fives. This wood may be shipped as far east as Florida, as far west as Japan. Grade lumber is used to make doors, window sashes, columns, and banister spindles. "That's old growth's niche," Walters says. "You take that high quality, that one-of-a-kind product, and use it in an application that demands an excellent appearance."

Isn't this a dead-end business? I ask. Won't all the old-growth trees be cut within a few decades? "There'll be plenty of it left," Walter replies, "but it'll be locked up for three percent of the population."

95

In just one sentence, he has managed to sum up much of the forest products industry's concern for the future. The government is seen as caving in to environmentalists and creating an artificial timber shortage. Some of the best old-growth timber on the face of the earth stands on the hillsides above the town of Darrington. Instead of selling it, the government has put much of the region's timber into national parks, wilderness areas, and now in spotted owl habitat areas. Some wilderness is fine, I'm told over and over by the people of timber country, but things have gone too far. It wasn't long ago that the sawmills could have all the wood they wanted. Now this timber town finds itself in an era of limits, limits seemingly set not by nature but by a political process. If more of the old-growth forest is protected from logging, it is argued that only an "elite" group of backpackers from the city will benefit.

Darrington, a town of a thousand souls nestled into the Cascade Mountains seventy-five miles northeast of Seattle, is truly a logging town. It is also a company town. Summit Timber, *the* company, dominates the local economy as much as the spectacular snow-covered peak of White Horse Mountain dominates the landscape. Summit employs 450 workers in this small town. Like most of the landless sawmills of western Washington and western Oregon, this company relies on federal timber sales to keep its saw blades turning. In the past, Summit and the smaller companies it has gobbled up over the years were able to obtain ample timber from the huge Mount Baker-Snoqualmie National Forest--usually from the woods near Darrington. Now, says Summit's logging chief, Walt Robinson, Jr., the company's "working circle" stretches north to the Canadian border, south nearly to Mount Rainier "and as far east as the timber goes." This summer, Robinson has trucks hauling logs over the mountains from a timber sale more than 200 miles away in the Wenatchee National Forest. Some townsfolk, dubious about the economics of that timber purchase, see it as a harbinger of tough times ahead. With the Mount Baker-Snoqualmie and Wenatchee national forests both planning reductions in their annual timber sales, independent sawmills soon will be fighting each other for a dwindling supply of logs.

It's quite a comedown for the Tarheels who migrated from western North Carolina to northwestern Washington in search of economic opportunity three-quarters of a century ago. They came here because they were promised jobs and there seemed little prospect of *this* timbered land running short of wood. The hardwoods of Appalachia had been largely cut over, while the timber industry was just hitting its stride among the giant Douglas firs of Washington. One of the first Carolinians to make the move was Dave Mallonee, who became superintendent of the Sauk River Lumber Company in Darrington. Mallonee was unhappy about depending on the "tramp loggers" who came and went as they pleased. Labor problems became a crisis in 1910 when white workers rounded up the twenty Japanese workers that one Darrington mill had just hired. Local ruffians put the Asians on a train and told them not to come back. Mallonee saw the solid family folk of Appalachia as the answer to his company's labor shortage. He sent word back to North Carolina offering a job to whoever would make the move across the country. Rail fare for workers and their families was part of the deal. More than fifty men, women, and children made the trip. Other waves of Carolinians crossed the country during the following two decades until Darrington was almost exclusively a Tarheel town. To this day, many Darrington folks speak with a drawl. The town even trades on its Southern heritage with a popular bluegrass music festival each summer.

Darrington's town elders today are the children of the men and women who migrated from North Carolina. But the frontier has been pushed back, both literally and figuratively. One who has seen the changes is Dewey Bryson. Born of Tarheel parents in a Skagit River logging camp to the north, Bryson grew up to do what was expected of every young Darrington man. He became a logger. Today, semi-retired, he sits on the patio of his Sauk Prairie rambler and recalls how the land has changed: "I can remember when we had a spar tree right about where that light pole is. The Sound Timber Company logged it. I remember this was all big old-growth timber here." Where his house now stands, he once had a small shake mill--"kind of a community affair" in which friends and relatives could split shakes and the profits when the logging camps were shut down for the winter. Houses with well-trimmed lawns now dot the countryside, along with pasture and small second-growth tree farms.

Like many loggers, Bryson went into business for himself as a gyppo, either contracting for a large mill or buying timber and selling it to a mill. When the federal government banned the export of raw logs two decades ago, gyppos lost a lucrative market and most sold out to large mills like Summit Timber. Bryson stayed in business, continuing to take on occasional logging jobs. He now works part-time as a mechanic at the Forest Service's Darrington Ranger District vehicle shop. The shop talk worries him: "I hear they're cutting back the sustained yield 30 percent next year. I don't know. I just hear the guys talking there. I can almost see it because they're turning more of the forest into parks and wilderness areas all the time, aren't they?"

The sustained-yield capacity is the Forest Service's calculation of how much timber can be cut annually on a continuing basis. Reductions in federal timber sales translate into layoffs at Summit Timber. The town's one big employer buys 85 percent of its timber from the Forest Service, the remainder from private owners and the state. Darrington folks always have had mixed feelings about the federal government. Even before Prohibition-era stills dotted the woods, there was grumbling about the government's decision to retain ownership of some forestland. Creation of the national forests, or forest reserves as they were known initially, was a sore spot from the first. When President Grover Cleveland expanded the forest reserve in 1897, one Darrington-area miner offered his opinion to the Arlington newspaper: "I

Top: A Summit Timber employee "bucks" a log into sections before it's yarded across the valley. Bottom: This seven-foot-diameter Douglas fir log snapped two cables as it was being yarded on a "salvage sale" on the Mount Baker-Snoqualmie National Forest.

believe that the reserve was created for the benefit of a few lily-fingered gentlemen who want a place to hunt and fish. I can imagine no other reason."

A few, like Barney Dowdle, who grew up in Darrington and became a forest economics professor at the University of Washington, still argue that there's no more justification for the government to own forests than to own farms or factories. No one--not even Dowdle--has any illusions, though, that the government will sell off or give away its forests. It's now a matter of making the best of a situation that won't go away.

I ask Bryson whether he would advise a young person today to plan a career in the timber industry. "He'd better get an education," the old logger says, "because, by golly, I don't know. Right now things don't really look too awful bright for Darrington. But we've said that before. We've gone through that time and time again. We've got upset and said Darrington will be a ghost town. But the population is the same. Some people may go on welfare but they'll pull through it somehow."

Towns like Darrington, almost totally dependent on the timber industry, are scattered through the Cascades, Olympics, and Oregon and Washington coast ranges. A few have become virtual ghost towns, and at least one company-owned town has been bulldozed into oblivion as the timber industry has moved on from that logged-over area. If Darrington's timber economy collapses, however, it won't be because the Forest Service scalped the entire countryside and then had to wait a century for a second-growth forest to grow to maturity. The Forest Service has made "community stability" a top priority. Added to the legal mandate that the forest be cut no faster than a new forest can be grown to replace it, the community stability policy would seem to give strong protection to a town like Darrington.

In reality, it doesn't work out that neatly. One of the reasons is that the Mount Baker-Snoqualmie National Forest's commercial land base--the land on which it may sell timber--keeps shrinking. It shrank in 1968, when the North Cascades National Park was carved out of the forest. It shrank again with the 1984 Washington Wilderness Act that enlarged the Glacier Peak Wilderness east of Darrington and created the Boulder River Wilderness to the south and west. In the late 1980s, the Forest Service prepared to remove from the timber base land on which soils were unstable, on which a new forest could not take hold within a few years, or which provided critical wildlife habitat. By failing to inventory those "unsuitable" lands earlier, the Forest Service had been overcutting its forests. Each reduction in the commercial land base reduced the national forest's capacity for sustained timber sales. Annual sales have dropped accordingly and the sawmills of Puget Sound have been forced to compete for a diminished supply of timber. Declining timber sales are viewed by many loggers and company executives as a breach of faith; homes have been built and sawmills constructed or modernized in reliance on government plans that ended up on the scrap heap.

Loggers aren't passively accepting what the government is dishing out.

When the Mount Baker-Snoqualmie National Forest managers held an "open house" at the Darrington Ranger District headquarters to discuss their proposed management plan, loggers turned out in force to protest it. As many as 100 logging trucks--many of them from the Summit Timber plant across the road--formed a noisy caravan. Large letters on the sides of the first dozen trucks spelled out their message: "ATTENTION - NEW FOREST PLAN - PLEASE DO NOT REDUCE TIMBER HARVEST LEVEL - SAVE TAX BASE - SAVE JOBS - SAVE DEPENDENT COMMUNITIES." Putting locals' feelings more plainly, the Backwoods Café put its own message on a reader board: "SPOTTED OWL STEW 99¢."

Truck convoys have become a favored industry tactic for garnering the kind of public attention that environmentalists gain with blockades of logging roads. Seven months after the Darrington cavalcade, more than 1,200 logging trucks roared through Grants Pass, Oregon, calling for more salvage sales in fire-damaged forests. Even more effective for its dramatic flair was the "Great Northwest Log Haul" organized after environmentalists' appeals of timber sales forced a Montana sawmill to shut its doors. Three hundred trucks from five states converged on the small town of Darby, Montana, with enough logs to get the mill back in operation. The log haul was capped by a rally in what one industrialist called "the Woodstock of the timber industry." Charlie Decker voiced the frustration that he and other loggers have been feeling: "The farmers get government aid, state help, and Willie Nelson concerts. All the timber industry gets is the Forest Service, workman's compensation, and the Sierra Club trying to put our picture in the post office."

The lack of respect that loggers receive from an environmentally conscious public troubles loggers as much as do declining federal timber sales. Whatever the merits of current logging practices may be, residents of Washington's growing cities and suburbs have little contact with the people of timber country. As the percentage of the population involved in the timber industry continues to fall, the public shows less and less patience with good *or* bad logging practices. Residents of outlying suburbs complain when second-growth forests are logged nearby. One of my neighbors in Seattle calls loggers and the Forest Service collectively "the bastards." It's not an uncommon attitude.

"Twenty years ago it was fun," Walt Robinson, Jr., says of the logging business. "I think any logger will say that. It was one of the most fun jobs you could imagine. It's not fun any more. I think anyone will tell you that." Beneath his gray hair combed straight back is a face remarkably free of lines and wrinkles for a man old enough to have fought in the Korean War. Robinson could retire more peacefully if the world hadn't changed so much in the past twenty years. Back then, when he and his partner Don Knowles were gyppo loggers, they didn't worry much about a log shortage. There wasn't so much paperwork when dealing with the government. Federal prosecutors weren't accusing timber companies of theft and bid rigging.

Above all, loggers hadn't yet become strangers in their own land. Robinson and Knowles sold out to Summit Timber in 1971. Today, Robinson runs G&D Logging, the logging arm of Summit. Knowles is one of his assistants.

Robinson was taken aback at a recent party when a city hall clerk took him to task for a Summit Timber clearcut on the steep hillside that's just across a pasture from downtown Darrington. He had a hard time convincing her that the gash in the hillside soon would be covered by a stand of young trees. The clearcut already had been replanted with 300 seedlings to the acre. After the party, the clerk gave Robinson a copy of a *Far Side* cartoon that depicts a group of trees surrounding a logger and taking away his chain saw. One of the trees says, "Let's cut him down and count his rings!" Robinson and the clerk remain on friendly terms, but he's disturbed by the lack of understanding some of his fellow townsfolk show of the timber industry. Two decades ago, a Darrington resident unhappy about clearcutting probably would have kept her opinion to herself.

"It used to be that almost everyone up here was a logger," Robinson recalls, "and almost everyone knew everyone else and almost everyone came from the same general area of North Carolina. I'm going back to the thirties. Up through the fifties, and probably the sixties, probably the biggest change that I can see is since all the independent operators went out of existence and it's one big company, there's not the civic pride that there used to be. A lot of hired help comes from outside town. There's more and more and more people retiring up here, or people that have jobs that they can commute to--say a professor that works three days a week, an airline pilot, there's a broker up here. Our circle is smaller than it was."

The newcomers are moving to small towns like Darrington because they love the countryside, not because of job opportunities. Some of them are making common cause with city dwellers upset over the logging of virgin forests. It galls Robinson that some environmentalists are trying to save forests they have never seen, that are far from hiking trails, and that have few giant old-growth trees. Most of the old-growth trees Summit cuts these days aren't the huge Douglas firs and cedars of days gone by, Robinson says. More typical is "half-rotten hemlock that's never had a hiker in it and in most cases doesn't exceed three feet in diameter."

It's true that most of the logs coming out of the woods these days are the smaller variety. But my visits to two Summit clearcuts showed me that big trees still are being pulled out. On a ridge above Goodman Creek, men were felling old-growth yellow cedars that were of large but not spectacular size. Along the creek, though, assistant bullbuck Henry Dickson told me that loggers had been busy a few days earlier with eight- and nine-foot red cedars down by the creek. Up on Prairie Mountain just east of Darrington, the Forest Service offered a "salvage sale" to take out all the trees dead or alive from an old-growth stand damaged by a windstorm. There, standing on the stump of a large cedar, I watched as the shovel operator struggled with a seven-foot Douglas fir so heavy it had snapped two yarding cables.

Robinson doesn't deny that big trees are being cut, but he wonders why some environmentalists consider that a sacrilege. Figuring an average lifespan of 500 years for old-growth Douglas fir, he calculates that perhaps fifty generations of trees have grown in the woods around Darrington since the last Ice Age. "That says to me the only thing that's unique about that tree we're looking at is that we're looking at it. It takes a lot of ego to put that importance on the tree just because we're looking at it. I know for any Earth First!er, that will put them out in orbit. But for each tree you have, there have been fifty standing there since yesterday."

The mobile home that serves as G&D Logging's office is separated from Summit Timber's modest headquarters building by two sawmills, a drying kiln, and sprawling log decks. In all, the complex covers about 200 acres. This company is small-time by the standards of, say, a Weyerhaeuser or a Georgia-Pacific. It's plenty big, though, by Darrington standards.

Gary Jones, president and CEO of Summit Timber, would like to concentrate his efforts on cutting costs and broadening markets. A neatly groomed man on the early side of middle age, he pulls out a business card that says something about his marketing priorities. One side of the card is printed in English, the other in Japanese. The day we talk, however, he is more concerned about "terrorism in the woods" than about the yen-versus-dollar exchange rate. A few days earlier, two saws were knocked out of commission when they sliced into some twenty-penny nails. Earth First!ers, taking a page from the book of the Wobblies during World War I, have been spiking old-growth trees in an effort to save them. Earth First! activists say their intention isn't to hurt mill workers--spikes like these can be lethal in a sawmill--but to keep loggers from cutting virgin forests in the first place. Normally, Earth First! notifies the Forest Service and/or the timber buyer that trees have been spiked. Jones says he doesn't know where these spiked logs came from. Logs are brought to the mill from many clearcuts, some logged by Summit or its contractors. Logs also are bought from other mills.

This wasn't the first "monkeywrenching" act affecting Summit. A week before the band saws ripped into spikes, two pieces of road-building equipment were vandalized north of Mount Baker. "There is no question this was done by experts," Jones says of the vandalism. "They cut every hose, every metal line of the injectors." He puts the damage at $30,000. Summit Timber has not stopped logging old growth in response to these fifth-column attacks; instead, it has beefed up nighttime security at logging sites.

If Earth First! were alone in trying to save old growth, it would represent nothing more than an inconvenience to the timber industry. But by the late 1980s, the nation's top environmental organizations had put the Northwest's ancient forests at the top of their agenda. The same kind of coalition that had saved the Grand Canyon and millions of acres of Alaskan wilderness was now working to put large tracts of ancient forests out of the reach of loggers.

Jones frets that "the hard-core environmentalists will never be satisfied until there's not another tree to be cut. . . . It would be a shame. This is one of the best timber-growing areas in the world. For us to take that resource or that crop and tie it up into an environmental issue or a recreation issue and a wilderness issue, it's a shame because they can coexist. There's no question they can coexist."

Environmentalists generally agree that there is room both for commercial timber production and preservation of natural forests. The debate isn't over the 18 million acres of managed second-growth forest in western Oregon and western Washington. Nor is it over the 1.1 million acres of mid-elevation forests protected in wilderness and national parks. Rather, it's over the fate of the 4 million acres of virgin forests that could be logged on Forest Service and BLM lands.

Where are the lines to be drawn? In the view of most loggers and foresters, the answer to that question grows out of their view of the forest resource. To them, trees are a crop to be harvested; to do otherwise is to be wasteful. Jones, calling himself "as much an environmentalist as anybody," casts the issue in stark terms when he confronts the views of environmental organizations: "I don't think that they realize the reason the Lord created the resources. He created the trees for us to use. Timber is a crop. If we manage it properly, we would be able to exist with it forever. Not to use it would be a waste of a material that we would never be able to reclaim."

Jones's view echoes God's directive in *Genesis* that mankind is to have dominion over the earth. Bob Spence, vice president of Seattle-based Pacific Lumber and Shipping, goes a step further when he suggests a moral imperative to put the old-growth forests to commercial use. If the forests aren't exploited, he contends, environmental costs will only be foisted onto the Third World, with far worse consequences. Whether it's logging or shoe-making, America has the know-how and the will to act in a manner more sensitive to the environment. Spence fumes about the "immorality" and "self-righteousness" of forcing polluting industries overseas.

Of Washington's eighteen million acres of commercial forest, Spence asks, "Can we maintain our eighteen million acres and create management practices that are pragmatic in terms of the industry's needs, the wildlife needs, water needs, forestry needs, recreation needs, visual needs? That gets down to the gut issue of the whole thing: are we as a people going to say that we can't do those things and are we going to remove the acreage and not allow ourselves to even try to discover whether we can manage for all those things? Are we going to be scared of our shadow?"

If only the government would give the forest-products industry a free hand in the old growth, it is often suggested, industry's problems would disappear. Not all of industry's problems are the result of federal policy, however. The market for forest products is subject to the same kinds of volatile market forces that affect every other industry, from agriculture to computers. In 1978, there seemed to be no limit to how much lumber could be sold or to its price. Five years later, the bubble had burst and timber companies went to Congress to escape from timber contracts signed before the collapse. After another five years, the market had rebounded and timber executives proclaimed that the most serious problem facing them was the looming timber shortage.

During the heady days of the late seventies, Summit Timber made plans to capitalize on the bright future. After decades of buying up smaller mills and concentrating operations in a single plant, the privately owned company was ready to make a leap to twenty-first-century technology. A new plant would be built to handle the smaller logs that represented a growing portion of the timber supply. With its laser and computer technology, the mill could produce greater volumes of lumber at a much lower unit cost. The $20 million plant was nearing completion in 1981 when the inflation-plagued housing market tumbled, and took the timber industry down with it. Summit was one of many independent sawmills left holding the bag in the form of long-term contracts to buy Forest Service timber. If the company logged the timber at the high prices bid before the crash, the owners would lose their shirts. If they defaulted on the contracts, there would be no revenue with which to service debt on their construction loans. Faced with operating losses on the one hand and 21-percent interest rates on the other, Jones went hat in hand to the lender, John Hancock Insurance. To pay its debt, Summit sold most of its 12,000 acres of land. The value of timberland at the time was as depressed as timber itself, so Summit took another financial beating on the land.

Still more drastic measures were required, though, to see the mill through. Workers agreed, after a three-week layoff, to two temporary wage cuts. The federal government provided relief as well, in the form of the Federal Timber Contract Payment Modification Act of 1984. Generally called the "buyback" law, this allowed timber buyers to escape from overpriced contracts. Like the stock market in the 1920s, the timber market of the 1970s had been characterized by ever-rising prices and a widespread faith that the boom would continue. Timber companies bid unprecedented prices for stumpage on the national forests. When the market collapsed in 1981, mills were saddled with long-term contracts that required them to buy timber at astronomical prices. Billions of board feet of timber had been sold in Oregon and Washington but weren't being cut. Stumpage prices on the national forests tell the story of the market collapse: average stumpage prices for old-growth Douglas fir plummeted from $475 per thousand board feet to $134 in just three years. Under the buyback law, no one bought anything back. Timber purchasers were allowed to pay the federal government a fee and escape from their contract obligations. Companies in Oregon and Washington paid $113 million to buy their way out of contracts totaling $2.3 billion.

Relief came too late for some, though. A hundred lumber and plywood mills closed their doors during the early eighties, and another 45 shut down between 1987 and 1989.

Summit Timber abandoned costly contracts for a whopping 80 million board feet of timber. It was then able to place more realistic bids on timber sales and resales. The company survived, but not without wounds. Workers had helped the company out again in 1986 by approving a labor contract that slashed the total wage and benefit package by nearly 20 percent. But in 1988, when lumber prices were back at high levels, workers took a hard line in contract negotiations. Workers struck for restoration of wages to previous levels. During that bitter dispute, strikebreakers were hung in effigy next to signs such as: "A GOOD SCAB IS A DEAD SCAB" and "SCAB SEASON OPENS SOON." As the strike dragged on from August into the new year, one Darrington woman refused to go to any family gatherings where she might meet her "ex-brother" who had crossed the picket line to return to work. The mill continued to work with nonunion labor.

Sawdust and the screams of metal meeting wood fill the air of Summit Timber's new mill. This much, at least, resembles any mill. There ends the similarity to mills of the past. The modern lighting and the brightly painted metal ladders and catwalks hint that this operation is different than the old-growth mill next door. Designed to handle "common logs" as small as five inches in diameter, this plant can crank out large volumes of lumber. Nearly everything is done automatically. Sawyers in high-tech control booths watch for problems and occasionally override the computers' judgments on how best to cut a log. Lumber is automatically sorted, checked for wane, trimmed, and sent to the kiln for curing. In a good eight-hour shift, the plant's sixteen workers can produce a mind-boggling 240,000 board feet of lumber. It takes nearly forty truckloads of timber to supply that shift. Enough lumber is produced to build two dozen homes.

With timber prices back up to 1970s levels, Summit's quad resaw mill is serving the company well. It represents the sort of modernization that some analysts believe necessary if the timber industry in Oregon and Washington is to compete successfully with the southeastern states, British Columbia, and, in the future, such timber producers as Chile and New Zealand. Modernization, coupled with employee wage concessions, has reduced Summit's manufacturing costs by 30 percent over eight years.

By reducing costs at the margin, it became profitable to truck logs 200 miles or more--something considered unthinkable in the industry just a few years earlier. "Our goal," explains Jones, "is to make our operation as cost-effective as we can, so if there's a stick of timber sold we have a good chance of buying it. Otherwise, we waste our investment and the community as a whole will suffer." While Summit was boosting its lumber production with fewer workers, it had more loggers bringing timber from the woods to the mills.

There's the rub. State-of-the-art sawmills like Summit's are indeed capable of producing lumber more cheaply--benefiting both investors and consumers. But the efficiency of a heavily capitalized mill can be realized only if there is an ample supply of logs (and if enough buyers want lumber). Summit committed itself to a high-volume operation at a time when the future log supply was beginning to cloud over. The forced divestiture of its own forestland left the company more dependent than ever on outside supplies. Private landowners were just liquidating the last of their old growth, and by the mid-eighties were cutting second growth at a rate that raised alarms about future supplies. Some private lands were being taken out of timber production, and there was little question that federal timber sales would continue to drop.

Summit's modernization and expansion would pay off only if the company could increase its share of federal timber sales. In the first few years of post-recession recovery, the strategy worked. Summit and other aggressive buyers filled the vacuum left by timber buyers who went under during the crash. As Summit reached farther for timber, bidding for the first time on sales in the Wenatchee National Forest, it brought in enough logs to keep the new plant busy two shifts a day. It was not clear that this could continue, however. The industry had not yet felt the impact of reduced timber sales that would result from planning under the National Forest Management Act.

Summit's heavy debt paled next to the debt-related tragedy in the California redwoods. To finance its buyout of the conservative, family-run Pacific Lumber Company, corporate raider Charles Hurwitz's Maxxam Group issued $795 million in "junk bonds." Burdened by that phenomenal debt, Maxxam dropped the former owners' sustained-yield policies and began liquidating its stands of 1,000-year-old redwoods at a frenzied pace. Loggers and mill workers joined environmentalists in protesting the rapid destruction of the forests. "They're just leveling everything. They're destroying the future, leaving nothing for the next generation," said one Pacific Lumber employee.

In western Washington and western Oregon, the problem was not leveraged buyouts. It was overcapacity in the mills. Even after the industry shakeout of the early 1980s, the industry's productive capacity outstripped the timber supply. In Puget Sound, just three mills dependent on federal timber sales--Summit, Mount Baker Plywood, and Portac--were capable of cutting every log the Mount Baker-Snoqualmie National Forest was selling. Yet they were competing against dozens of other mills for the limited supply.

Another industry shakeout was expected to come in the 1990s. No one could say with certainty which of the independent sawmills would survive the coming years. Would it be those like Summit that invested heavily in greater productivity? Or would it be those like Bruce Engel's Portland-based WTD Industries, that took advantage of the recession to buy twenty failing plywood and lumber mills at bargain-basement prices? WTD lacked Summit's technological sophistication, but it carried only a modest debt burden.

Throughout the industry, in mills like WTD as well as Summit, more lumber was being produced by fewer workers than ever before. Oregon and Washington lumber production set a record in 1986, but the number of workers required to produce the lumber dropped from 136,500 to 100,600 over an eight-year period. Some mills improved efficiency by adding work shifts or by pushing workers harder. Others used automation. Weyerhaeuser coupled its new profit-sharing program with wage cuts and a streamlined work force. The company restructured some plants to be "run by essentially unsupervised hourly crews," in the words of a senior Weyerhaeuser executive. The big companies cut costs further by contracting out more work to gyppo loggers. The downward trend in forest-products employment was expected to continue. The number of timber-related jobs in western Oregon and western Washington declined from 119,000 in 1970 to 103,000 in 1982. A Forest Service economist predicted that employment would plummet to 64,000 by the year 2000. That projection was based primarily on increasing productivity and plummeting timber cuts on the depleted private lands.

A timber industry analysis in 1988 forecast a loss of just under 5,000 timber-related jobs in Oregon and Washington if the Forest Service implemented its draft management plans for nineteen national forests in the two states. Those plans worked out to a 21-percent cutback in timber sales, according to the industry analysis.

One remarkable fact stood out amidst all the dry numbers. While timber executives were proclaiming a crisis on the public lands and berating the Forest Service for jeopardizing 5,000 jobs, those executives were offering no apologies for the loss of 55,000 jobs due to the depletion of the industrial forests and automation in the mills. Massive layoffs to make the industry more competitive were deemed perfectly acceptable. More modest layoffs to save an ecosystem were branded "catastrophic."

A 1988 study by The Wilderness Society, using a Forest Service computer model, projected nearly a 50 percent drop in timber industry employment by the year 2030 even if national forest timber sales remained constant. The decline resulted from additional productivity increases and from reduced timber harvests on the depleted industrial forests. If logging on the national forests were reduced by 25 percent to preserve the most ecologically viable old-growth forests, the report forecast an additional 6-percent reduction in the region's total timber harvest. That would translate into 2,300 jobs lost in addition to the 8,200 that would be lost anyway.

A little less than 200 miles south and west of Darrington is Grays Harbor, the log export capital of the West Coast. There, on a public dock, the *Ocean Master* lies tied up, her deck piled high with logs. Tomorrow, she will set sail for Pusan with a load of 3.5 million board feet of timber. Logs that only a few weeks ago were trees in an Olympic Peninsula forest soon will be building

Left: Longshoremen stay busy, but mill jobs are exported along with unprocessed logs. Bottom: Asia-bound logs account for one-fourth of the logs cut in the Pacific Northwest.

Korean homes and shops. The docks and log-storage yards at Aberdeen are so crammed full of logs that Karl Wallin, the harried operations director for the Port of Grays Harbor, says, "We're bursting at the seams." Acre upon acre contains neatly stacked logs, everything from pole-sized second-growth thinnings to the elephantine trunks of old-growth rain forest trees. Japanese log purchases are picking up again, so the congestion soon will be relieved a bit.

The slackening of Japanese timber purchases turned out to be only a temporary lull in the booming log-export business. Log shipments from Oregon and Washington to the growing Asian markets soared above three billion board feet in 1987, higher than any previous year except 1979. Exports hit a new high in 1988. Ships made 229 port calls at public and private docks in Grays Harbor to pick up logs from the Olympic Peninsula. More than one-third of Washington's timber harvest went overseas without any processing in domestic mills. This was good news for the exporters--notably Weyerhaeuser, ITT Rayonier, and the state Department of Natural Resources--who received top dollar for export logs. It wasn't good news for workers in local mills that can't get enough timber. Nor was it good news for those worried about the fate of the Northwest's last stands of old growth.

Across the road from the Aberdeen export yard, where cranes are lifting huge payloads of logs, looms the darkened outline of a shut-down sawmill. "If this is not ludicrous, I don't know what is," sighs Chuck Sisco, a National Audubon Society biologist, as he watches the *Ocean Master* prepare for departure. "If these logs are over here, they sure could go through this mill. Can you imagine how many jobs are sitting out here? It just blows me away." Sisco believes that in allowing the export of raw materials, the nation is behaving like a Third World country that lacks the industrial capacity to process its own natural resources. In his view, log exports not only cost mill workers jobs but leave local mills more dependent on federal sales of old-growth timber.

Under the ghostly yellow glow of sodium-vapor lights, we drive past acres of logs ready for shipment to Japan, China, Korea, and Taiwan. On our way out, past the storage yard, we see a stack of large western red cedar logs. Next to this stack--which could fill a good-sized freight train--lie the crown jewels of this storage area: the bases of three immense cedars. We step out of our car into a driving rain for a closer look at these giants. Each is about eight feet in diameter.

With us on this trip is Elliott Norse, senior ecologist for The Wilderness Society. We make a quick estimate of the age at which one of the trees was cut down. The annual growth rings are tightly spaced, between twelve and twenty to the inch. Multiplying that spacing by the four feet from center to edge, we estimate the age at around 750 years. That's a conservative guess, because the outermost edge of this ancient's base has been sawn off.

"It's very possible," Norse observes, "that we are looking at a tree that took root in the time of William the Conquerer."

This log, worth thousands of dollars, presumably will not be shipped overseas. The cedar in this Weyerhaeuser storage yard is sold on the spot market, but not to foreign buyers. Federal law forbids the export of unprocessed western red cedar. Weyerhaeuser, with $1.5 billion worth of exports in 1987, continues to be the nation's leading seller of wood products overseas. Half of its exports go to Japan. The company's shipments of raw logs from Grays Harbor have reached such proportions that the two berths at its Aberdeen export dock can't handle all the traffic. Weyerhaeuser and ITT Rayonier export more logs over their Grays Harbor docks than are harvested on the Olympic and Mount Baker-Snoqualmie national forests combined.

Japan is the major buyer of "clear," high-quality, old-growth logs. The Chinese and Koreans, more concerned about price than the aesthetics of the wood they buy, generally import cheaper second growth. The willingness of Japanese trading companies to pay top dollar for old growth--a whopping $100 more per thousand board feet than the American runner-up in one public auction--has run up stumpage prices and left independent mill owners fuming. If local mills matched the Japanese prices, they insist, they would lose money when they resold the wood in the form of lumber. The fall of the dollar against the yen during the late 1980s further reduced the ability of domestic mills to compete for logs.

Domestic mills might be priced out of the game altogether were it not for federal restrictions on the export of logs from federal lands. The secretary of agriculture and secretary of interior issued orders in 1968 limiting the amount of federal timber that could be exported as raw logs. That ban was incorporated into law the following year. The law was tightened, at least ostensibly, in 1974 to ban the export of federal logs altogether. Exports of lumber and other forest products are unregulated; after a log undergoes "primary processing" in a domestic mill, the wood may be exported. The law contains two big loopholes. Mills that had a history of exporting logs could continue to export logs from private lands and use federal timber in their mills to substitute for the amount exported. Those "direct substitutions" could not exceed 110 percent of their historic export levels. Fifty-one buyers of federal timber would be allowed to export nearly half a billion board feet of timber from western Washington, western Oregon, and northern California. The other loophole is what's called "third-party substitutions." If a timber buyer buys federal timber, then resells the logs to a friend, the friend may legally substitute those logs for exports of private timber. The General Accounting Office estimated in 1985 that 100 million board feet were being exported through third-party substitutions. That may not be a large number relative to total federal timber sales, but it represents a significant volume of timber for mills in southwest Washington, where nine-tenths of third-party substitutions take place.

The loopholes in federal law concerning direct and third-party substitutions pale next to exports of logs from private and state lands. Washington, the leading exporter, sent 2.6 billion board feet, or 37 percent of the state's harvest, overseas in 1987. The export trade is most pronounced when it comes to timber from state-owned land. At least two-thirds of the timber from Washington state sales goes straight to the export docks without passing through local mills. It is a situation that has put independent mill owners at odds with the state Department of Natural Resources and with large corporations. Those small, landless mills that depend on public timber sales have spent years fighting for a ban on the export of state timber.

The small mills' protectionist campaign scored victories in Alaska, Idaho, Oregon, and California--all of which enacted export restrictions. Then the U.S. Supreme Court in 1984 struck down the laws as a violation of the commerce clause of the Constitution. The high court left one avenue open to the small mills: Congress could enact enabling legislation that would authorize state-by-state restrictions. In 1987, Rep. Peter DeFazio of Oregon introduced enabling legislation. The bill was bottled up in the House Foreign Affairs Committee, with DeFazio blaming its defeat on a last-minute lobbying effort by big timber companies and the Reagan administration. "Their footprints were all over this," he fumed after the bill died.

By 1989, when court rulings placed federal timber sales in jeopardy, public pressure mushroomed in favor of controlling exports. DeFazio and Oregon senator Bob Packwood introduced bills authorizing bans on export of timber from state land. Oregon voters overwhelmingly passed a constitutional ammendment intended to restrict exports. Politicians in Washington began to explore, fitfully, the possibility of setting aside a portion of state-sold timber for domestic milling. Those efforts were frustrated by Public Lands Commissioner Brian Boyle's objection that export restrictions would mean lower timber prices and reduced revenues for the land-trust beneficiaries. Boyle didn't mention that while exports meant more income for his own Department of Natural Resources, they meant less income for the state general fund because of lost sawmill jobs.

Small sawmills pushed for restrictions in hopes of acquiring timber more cheaply and selling finished products to Japan. The big exporters like Weyerhaeuser and ITT Rayonier opposed restrictions. The export debate was a textbook case of competing political interests. The industry giants' interests were so different from those of the small mills that some insiders noted that it was a misnomer to speak of "the timber industry." In fact, there were two timber industries: one, the capital-rich, land-rich segment that held most of the cards; the other, the independent mills buffeted about by the decisions of the big companies and the government. This situation has not changed and is not likely to for the foreseeable future.

Environmentalists weren't sure how to address log exports, so they initially stayed on the sidelines. Some, like Audubon's Chuck Sisco, saw the issue as central to forest preservation. The public should be up in arms, he argued, over the Washington Department of Natural Resources' logging of its last old-growth forests--particularly when jobs were being exported along with logs. But the issue was tricky for environmentalists. How would the forest benefit if a 500-year-old Douglas fir was sliced up in a mill in Aberdeen rather than in Tokyo? But when court injunctions blocked many old-growth timber sales in early 1989, environmentalists pushed for export restrictions as a way of heading off the backlash against their legal victory.

Some activists suggested that raw log exports be taxed. The tax revenues would be used to fund expansion of the national forests and expand the supply of second-growth timber available to independent sawmills. Another possibility was use of tax incentives to induce industrial landowners to send their logs to local rather than foreign mills. The resulting boost to domestic mills could be coupled with increased protection of old-growth forests. Those ideas had two advantages over a flat-out ban on export of state timber. One was that they would use price signals rather than the regulatory process to change corporate behavior. The other, and more important, advantage of tax measures was that they would address the export of logs from private lands. Private-log exports were considerably larger than state-log exports.

The Bush administration's misguided proposal to allow the export of logs from the national forests gained no support in the Northwest. Opponents said federal revenues would be reduced if 2,000 mill workers were thrown out of work as predicted by the Congressional Research Service. The impending log shortage was fueling public support for more, not fewer, restrictions on exports.

Timber companies were selling as many raw logs as they could because premium export prices made it more profitable to ship the logs abroad than to mill them at home. One securities analyst estimated in 1979 that Weyerhaeuser earned a whopping 62-percent profit on log exports, compared to a 10-percent profit making lumber. If those figures were even remotely close to reality, a timber manager would be remiss in his duties to shareholders if he put millworkers' jobs ahead of the immense profits to be made in the export trade. Weyerhaeuser announced in May 1988 that it was closing its seventy-year-old old-growth sawmill in Snoqualmie, Washington, because there weren't enough large logs to keep the mill busy. Two months after that announcement, I watched workers at Weyerhaeuser's export dock in Tacoma debark and sort logs for export to Japan. Next to the water were old-growth logs, some of whose age foreman Bill Roberts placed in excess of 300 years. "That's our high grade here--one of the best," he said, kicking a four-and-a-half footer.

A truly international company, Weyerhaeuser opposed trade restrictions of any kind. With more than seven million acres of timberland under long-term lease in Canada and more mills in British Columbia than in Washington, the world's largest forest-products company ships products into the United

States as well as out. "We're free traders," said Weyerhaeuser spokesman Tom Ambrose. The trade associations representing big timber portrayed the drive for export restrictions as a cynical grab by small, inefficient mills for cheap public timber. "Some of them, if you gave them free stumpage, they would still be in trouble," said Stu Bledsoe, director of the Washington Forest Protection Association, a few months before his death in 1988.

Some free-trade advocates funded by industry even made the dubious argument that the Northwest would suffer a net job loss if logs from state lands had to be milled locally before being exported. A 1981 study commissioned by the Washington Citizens for Free Trade acknowledged that sawmills would pick up new jobs but that the gain would be more than offset by decreases in state timber-sale revenues. The most "realistic" scenario, according to the lobbying group funded by the big timber companies, would be a net loss of 3,300 jobs in Washington. Northwest Independent Forest Manufacturers (NIFM), representing independent mill owners, offered a very different analysis. If 600 million board feet of state timber were milled at home rather than in offshore mills, the group calculated, 9,000 new jobs would be created in Washington.

As it turned out, the log trade wasn't as "free" as it once appeared. The U.S. Commerce Department in 1989 accused Japan of a variety of unfair trade practices to protect its sawmills. Could the United States turn the situation around, or was it helpless to escape its position as a mere supplier of raw materials? NIFM pointed to Indonesia and Canada as evidence that the United States could be more than a raw-material supplier. When Indonesia phased out its huge log exports to Japan, it built 126 plywood plants. Indonesia jumped from a 1-percent share of the Japanese plywood market to an 18-percent share between 1984 and 1987. Canada, allowing only the export of logs declared "surplus" to domestic mills, had become the world's largest lumber exporter. Many U.S. mills would have to change their production methods in order to satisfy Japanese homebuilders accustomed to placing custom lumber orders with small, local mills. Some American mills are already producing custom lumber, and NIFM president Gus Kuehne contended that modern mills "can cut to foreign specifications with the flip of a few switches." He reported that some mill owners had orders from overseas customers but had to shut down because they couldn't get enough logs.

About the only thing large and small producers could agree on was the value of promoting finished products on export markets. Some industrialists were doing that aggressively. At Vanport Manufacturing in Boring, Oregon, owner Adolf Hertrich boosted sales by millions after he built a Japanese guest house at his plant and began turning out lumber in the metric dimensions favored by Japanese builders. At the new Pacific Veneer plant in Aberdeen, Weyerhaeuser and the Bank of Tokyo teamed up to produce plywood veneer to Japanese specifications. The Evergreen Partnership marketing consortium hoped to pique Japanese interest in American lumber and building methods by building a 170-home "Washington Village" as part of a planned city near Kobe. Pointing to Japan's 1.2 million annual housing starts, Evergreen Partnership executive director Greg Shellberg explained, "What we're really after is getting a share of those 1.2 million units." The effort to sell more finished products on Asian markets has been a success story. Oregon and Washington mills tripled the value of their softwood lumber exports to more than half a billion dollars between 1977 and 1987. Despite that impressive growth, sales of finished products continued to be overshadowed by the much larger trade in raw logs.

The irony of the situation was inescapable. While the forest-products industry blamed the Forest Service for selling too little timber, kingpins of the industry were exporting a volume of raw logs almost equal to the entire timber output of the national forests in the Douglas fir region. No wonder sawmills were scrambling for timber. And no wonder the timber industry was so eager to cut all the old growth it could sink its saws into.

Some loggers have been particularly innovative in their efforts to get enough wood. The cedar rats of Darrington's Select Shake are scrambling through the woods looking at half-hidden logs.

"Hey, Mike! There's a nice one!"

Mike Presnell makes his way through the old-growth forest to look at the log. It's nearly six feet in diameter, a cedar that's been on the ground long enough to be covered with moss and to have a fifty-year-old hemlock growing from it. Most of the wood in this nurse log is surprisingly solid.

Presnell's boss, Mike Roberts, examines the log. "Oh man, what a piece!" he exclaims. "It's a beautiful piece--not a knot in it."

The men walk excitedly through the woods, examining all the large cedar logs and snags. The company Roberts owns, Select Shake, is "prelogging" the Iron Mountain timber sale. Summit Timber, the high bidder, will clearcut 187 acres of virgin forest. But before Summit's loggers rev up their chain saws, Roberts' loggers will salvage as much good cedar as they can find. It's good business for everyone. For Roberts, old-growth cedar is the raw material that keeps his shake mill running. Summit benefits from the removal of large logs that otherwise could cause trees to shatter when they're felled. The Forest Service wins because it gets additional revenue. From one five-foot cedar, Presnell and his crew have cut eight cords of wood and tied them into bundles that will be yarded by helicopter. "That one down tree brought the government $2,000," Presnell tells me.

Another sound log lies a stone's throw away. For two days, Roberts's cedar rats have been sawing it into two-foot lengths, splitting those lengths into smaller blocks, and then bundling them with stout ropes. The best wood will be split into shakes, the second-best will be sawn into shingles. Defective wood is left on the ground. Each day Roberts's eight loggers stay on this job, they're generating $12,000 of revenue for the government. It's a slow, labor-intensive process to remove a valuable and increasingly rare commodity.

These woods on Iron Mountain are a shake producer's dream. The dominant red cedar is unusually large and abundant. The understory is predominantly hemlock and silver fir. One of the live cedars that will be left for Summit Timber's clearcutters is easily eight feet in diameter at breast height. Growing out of its side is a huge "school marm," or candelabralike branch. These woods are perfect for cedar scavengers because the giant cedars are doing what old trees do. One by one, they're dying off, making way for their smaller successors.

The volume of salvable cedar is enormous. Out of one promising log grows a foot-thick silver fir whose age Presnell puts at 100 years. Not all of the cedar is merchantable. One member of the logging crew looks at a log and sputters, "This is garbage." Another log--"It's a punkin!" Roberts says of the whopper--is so deeply covered with moss that it takes the scavengers a while to determine that it's a cedar, not a Douglas fir. They're dubious whether it will yield shake-quality wood.

It's taken some salesmanship to persuade the Forest Service to allow prelogging like this. "We need to be the eyes for the Forest Service," says Roberts. "We need to be the guys who identify opportunities. They don't have enough guys in the field and the guys they do have aren't motivated by the same thing we are." Timber sale officers in the Mount Baker-Snoqualmie National Forest are willing to go along with prelogging--*if* someone like Roberts is out there pushing for it. Prelogging fits the Forest Service's idea that all merchantable material should be removed from a clearcut. Roberts is keenly aware that some forest ecologists are lobbying the agency to replace current logging practices with methods that would leave behind a substantial number of live trees, snags, and logs. That could spell doom for the prelogging business.

Where an ecologist might see rich plant and animal habitat, a logger sees something else. "It's defective, it's diseased, it's overripe," Roberts says of this stand. "It's a fire hazard and it's close to town." He and his employees are eager to show me that this ancient forest offers meager wildlife habitat. We cross a deer trail, and the loggers are quick to point out that it hasn't been used in many months. "Where do you find deer feeding?" Presnell asks. "Not in the timber. It's in the clearcut." We come to a hollow cedar snag in whose bark a hemlock has taken root forty feet above the ground. Roberts pokes his head into the cavity and, seeing no sign of animal life, remarks that this is exactly where one would expect to find something. I wonder aloud about the deep pile of flaked wood in the hollow; the loggers attribute it to the wind, not to burrowing or foraging animals.

Whatever the ecological merits of salvaging and clearcutting old growth, Roberts's thriving business is the embodiment of entrepreneurship. Not unlike the salamander that makes its home in large rotting logs, he has found a special niche in the old-growth forest--and in the timber economy. Using as capital his profit-sharing nest egg from the twelve years he worked at Summit Timber, Roberts joined with two partners to buy a shut-down shake mill at a time when other mills were going out of business. It was a high-risk venture that could easily be dismissed as sheer folly. All the mills were having trouble getting enough cedar. Some were beginning to pin their hopes on a new preservative process that might allow hemlock to be split into shakes. Roberts wasn't plugged into the kind of old-boy network through which some shake mills obtain their cedar from the large lumber mills. Could they make a go of it on the basis of sheer guts and determination?

Before the end of its first year, the company ran into serious troubles. One of the partners backed out and costs piled up more rapidly than revenues. Debts rose to $200,000, and the two remaining partners filed for protection under Chapter 11 of the bankruptcy code. But they didn't give up. "We busted ass all season and we pulled out," Roberts says with pride. The debts were paid off and the company came out of Chapter 11. This summer, Select Shake has eight crews out in the woods. Counting loggers and mill workers, the company payroll totals forty. In a town the size of Darrington, this represents an important new infusion of economic vitality. Mike Roberts hopes to bring more stability to the enterprise next year by adding a specialty old-growth Douglas fir operation. But he's troubled over his winter's cedar supply, which was to have come from the Gold Coast timber sale. The sale has been held up due to an appeal by a local Audubon Society chapter.

The financial roller coaster that a small mill rides is similar to the rides that gyppo loggers have taken over the years. Gyppos--the small, independent operators--came into existence in the 1940s with the advent of truck logging. Up through the sixties, many entrepreneurs succeeded in being their own bosses. In those days, they could buy timber from the Forest Service and resell it to foreign buyers. When the Forest Service banned the export of raw logs, the gyppos were left at the mercy of the sawmills. Many, if not most, sold out to the mills. Some of the survivors today stay in business by virtue of another government regulation.

The Forest Service guarantees that a percentage of its timber sales, based on historic patterns, will go to "small businesses." A small business may have no more than 500 employees. For a mill to hire a five hundred and first employee--and lose the small business preference--would be suicide. Summit Timber, after decades of expansion, has safely stabilized its work force at around 450 employees. Summit remains a "small business" by hiring gyppo loggers to log and haul much of the company's timber. Pacific Lumber and Shipping, which runs its $200 million business out of Seattle's presigious Rainier Bank Tower, also is a "small business." It has only 480 employees.

There are two ways a company like Summit Timber or Pacific Lumber can keep expanding and still take advantage of the government's relatively cheap old-growth timber. It can contract out more of its logging and it can automate its production lines. Summit has done both.

Loggers, far more than mill workers, are a rugged, individualistic lot. Given half a chance, they'll strike out on their own. And that, as often as not, means financial insecurity. On a foggy, rainy morning near Mount Pilchuck, a gyppo logger told me how he had used his log loader, or "shovel," the previous day to smash a twelve-foot-diameter cedar log that was too big to fit on a truck. Asked if he had any reservations about his work, he answered, "You want to hear our motto? 'Over the rivers and through the hills, we rape the land to pay our bills.'" He paused after saying that, realizing that his joke might be taken more seriously than he intended.

Some loggers have taken to calling themselves an endangered species. What right, they ask, do environmentalists have to protect the spotted owl or the pine marten at their expense? "It's plain and simple," opines Mike Presnell. "What's more natural than us living here and doing what we do?"

While Select Shake builds a new mill in Darrington, the company is operating on a leased property in Swede Heaven, a small settlement outside town. The small building, open to the air, is a beehive of activity. One man slices shingles on a circular saw that was built in 1916 but is as efficient as ever. Another holds blocks of higher-grade cedar while a mechanical splitter busts them into half-inch-thick shakes. The truly frightening part of the operation is to watch a sawyer resaw each reactangular shake in half. He stands the shake on its side and pushes one corner through a screaming band saw. When the leading edge of the shake clears the blade, he reaches past the blade to grab it. One hand pulling away from the blade, the other pushing toward it, he guides the shake so that the saw blade slices neatly through the opposite corner. Sawdust showers over the sawyer, who spends only a few brief seconds cutting and trimming each shake. The two shake sawyers and one shingle sawyer quickly slice off any imperfections before throwing the finished product down a chute. A team of packers bundle the shakes and shingles, then stack them on pallets. One packer wears a strikingly environmentalist T-shirt: "Keep the Magic--SAVE THE SKAGIT."

One image from the shake mill won't leave my mind. That's the sight of the sawyer's hands dancing around the band saw. I'm awed by his proficiency--and by his ten intact fingers. For him, the hazards are just part of the job. With enough agility and alertness, he can turn out thousands of shakes in a day and keep his fingers whole. But all of his agililty--and perhaps even the business agility of Mike Roberts--won't be enough to keep this job and this industry alive.

This mill will stay in business only as long as the Forest Service keeps selling old-growth timber. In time, of course, there won't be any more old growth to sell. Neither the Forest Service nor the forest-products industry sees fit to grow cedar or other trees on the kind of long-term rotations that would provide a continuing supply of high-quality timber for specialty use.

Federal timber management aims to keep specialty mills alive in the short run. It guarantees that they will die out in the long run.

6

Whose Woods These Are

The frontier is still being pushed back in the Pacific Northwest. Yet-unvanquished parts of the natural landscape surrender each day to human will. In few places are the scars as fresh and as extensive as in the rugged mountains southeast of Mount Olympus. Roughly 100,000 acres of federal land has been clearcut over the past four decades--a number whose meaning can be understood only by looking at the land.

In the Olympic National Forest's Shelton Ranger District, the natural forest has been cut, bit by bit, until there's almost none left. The bleakness of cutover land stretches from one steep ridge to the next, as far as the eye can see through a distant haze. Small, isolated stands of old growth only heighten the feeling of desolation. They serve as a reminder of what has been lost.

This wounded landscape brings to mind the words of two former chiefs of the Forest Service. Edward P. Cliff likened old-growth clearcuts to an urban renewal project in which destruction is "a necessary violent prelude" to progress. William B. Greeley understood the shock of the passerby who sees a landscape transformed into "a fitting approach to Dante's *Inferno*." The devastation on the Shelton Cooperative Sustained-Yield Unit is so extensive that environmentalists call it the "sustained-rape unit."

Many in the timber industry and not a few economists feel that if other federally managed forests in the Douglas fir region were properly managed, they would look more like this. The Shelton experience demonstrates both the risks and rewards in departing from the Forest Service's standard timber-sales practices. It also points out the extent to which management of the last publicly owned old growth is influenced by the management of private lands.

Under a unique 100-year agreement with the Simpson Timber Company, the Forest Service agreed to merge a block of its land with Simpson land into a single "cooperative unit" for purposes of calculating the land's capacity to produce timber on a sustained basis. The Sustained-Yield Forest Management Act of 1944, which authorized the deal, was engineered by industrial landowners who were looking for a way of obtaining more timber after they liquidated their old growth. Before World War II, industry lobbyists discouraged federal timber sales in order to maintain timber prices and encourage reforestation of private lands. But as the war ended, the industry for the first time began to face the prospect of future log shortages. The industrial lands had been logged off so fast that the new tree farms wouldn't reach harvestable age before the supply of privately owned old growth ran out. The timber industry had plenty of very young timber and a dwindling supply of very old virgin timber. In between was a gap in the age classes that could meet the industry's needs in the future. The industry set its sights on the national forests.

Two days after the Sustained-Yield Act was passed, Simpson Timber applied to the Forest Service to create the nation's first (and, as it turned out, the only) cooperative sustained-yield unit. The deal signed in 1946 gave the timber company an assured supply of logs for its mills in Shelton and McCleary. Simpson gained exclusive access to the federal timber in the sustained-yield unit. That timber kept the mills busy after Simpson's 237,000 acres were depleted. By the mid-1980s, when the old growth on the national forest was almost gone, Simpson began harvesting second-growth timber on its own property. The company's timber cut in the Olympic National Forest dropped from 96 million board feet in 1984 to less than one million in 1987.

Had timber sales on the Shelton Ranger District been handled as they are on other national forest ranger districts, it would have taken close to a century to exhaust the old-growth forests. Instead, the woods were leveled in a little over forty years. To understand why the pace of cutting has been so much heavier on the Shelton district than elsewhere, it's necessary to understand something about the Forest Service's usual practices.

Timber sales on the Forest Service's 191 million acres are based on what the agency calls the "long-term sustained yield capacity" of the forest. In somewhat simplified terms, this means a national forest's timber sales won't exceed the annual growth rate of its timber. Sales will fluctuate somewhat from year to year, but over the long run they are to remain constant. A national forest's timber sales in 1995, theoretically, will be about the same as its sales in 2035. Timber is offered for sale with little regard to market demand and even less regard to the forestry practices of private landowners.

This conservative approach to forest management dates back to the agency's first chief, Gifford Pinchot. Trained at *l'École nationale forestière* in France, Pinchot was one of the first Americans--and by far the most influential--to bring European notions of sustained-yield forestry to this continent. "Forestry is Tree Farming," Pinchot wrote at the end of his career:

> Forestry is handling trees so that one crop follows another. To grow trees as a crop is Forestry.
>
> Trees may be grown as a crop just as corn may be grown as a crop. The farmer gets crop after crop of corn, oats, wheat, cotton, tobacco, and hay from his farm. The forester gets crop after crop of logs, cordwood, shingles, poles, or railroad ties from his forest, and even some return from regulated grazing.

The national forests were established with two initial goals in mind: protecting watersheds and managing the public lands in a more responsible way than the cut-and-run that was standard practice in the nineteenth century. Pinchot's vision, far more advanced than that of his contemporaries in the timber industry, was that the national forests would be operated as model tree farms. Seed trees would be left on cutover areas, and the rate of cutting would be limited. In a sense, the Forest Service saw its mission as one of saving the timber industry from itself. The national forests were one place where sustained-yield forestry would be practiced, regardless what happened on private lands. They would produce a dependable and stable timber supply.

Coupled with his message of a brighter future through sustained-yield forestry, Pinchot brought a warning. President Theodore Roosevelt, influenced by the man he put in charge of the nation's forest, told Congress in 1905 that a "timber famine" was inevitable in the face of large-scale logging without reforestation. Five years later, Pinchot declared that the crisis had arrived: "The United States has already crossed the verge of a timber famine so severe that its blighting effects will be felt in every household in the land." Pinchot wasn't the first to perceive disaster on the horizon. In the late nineteenth century and the early years of this century, virgin forests were being cleared so rapidly that fear of shortages grew widespread. Former Secretary of the Interior Carl Schurz warned in 1889 that another twenty-five years of forest destruction would leave the United States "as completely stripped of their forests as Asia Minor is today."

A full century after Schurz's warning of impending disaster, America hasn't been deforested. Three-quarters of a century after we crossed the verge of Pinchot's "timber famine," a famine has yet to be seen. Where did the doomsday predictions go wrong?

Pinchot, for all his appreciation of silviculture, failed to appreciate nature's ability to regenerate forests. That ability was apparent to none other than Frederick Weyerhaeuser. The world's most successful lumberman declared in 1910, "There is no reason to think the timber supply will not hold out indefinitely." There were other factors that put the lie to the doomsday scenarios. After more than three centuries of deforestation, many farmlands were taken out of food crop production and reforested.

And there was a type of conservation that Pinchot failed to consider: reduced consumption of an increasingly costly resource. The railroads, chief among the industrial users of wood, were alarmed by foresters' timber-famine predictions at the turn of the century. At first, railroads began growing their own trees to bridge the expected timber supply gap. In the end, they found a more cost-effective solution to the problem. The railroads simply found ways of using less wood. Steel and concrete replaced wood in many uses. Tie plates distributed the load on ties, extending their lives. Use of wood preservatives allowed cheaper woods to be used, with less need for replacement.

"The experience of the railroads as consumers of forest products provides a window on the American forest economy and puts the old predictions of timber famine and commercial disaster into a new perspective," Sherry H. Olson observed in her excellent study of railroads and timber. "The crucial error lay in an important confusion: physical facts of production and consumption were confused with economic facts of supply and demand."

Demand, then, can be understood in more than one way. Wood is not like a rare metal used only in high-performance military aircraft, for which demand can be defined or quantified exactly. In a market economy, demand for a commodity like wood fluctuates on the basis of supply, price, and availability of substitute products. No one speaks seriously these days about a "timber famine" in the United States. Tree growth today far exceeds timber harvest (though not in the Northwest, where old growth is still being logged). The nation's timber inventories are being built up at a pace probably unprecedented in history. This despite the fact that lumber production and consumption are also at record levels.

From a national perspective, sustained-yield forestry is working quite well in meeting demand. Globally, the situation is similar. Even while the tropical rain forests are being recklessly depleted and firewood-gathering in the Sudan converts forest into desert, the world timber market is about to be flooded with Monterey pine from the plantations of Chile and New Zealand. The northern forests of the Soviet Union and Canada remain largely untapped. As far as the consumer is concerned, the now-global wood market will provide ample supplies into the foreseeable future.

When executives in the forest-products industry of the Pacific Northwest talk about an impending timber shortage, the shortage must be placed in context. A shortage of sorts *is* approaching for a number of reasons. There aren't enough logs available at competitive prices to keep independent mills operating at capacity. There is enough timber, however, to supply this nation's need for wood and paper products. There are even enough logs to allow the big timber companies to ship billions of board feet to Asia every year. It's largely a question of who sells the logs, to whom, and at what price.

A crisis erupted over forest management in early 1989 when the Olympic National Forest revealed that its timber sales would be reduced by one-half. The main reason for the reduction was that the timber sales staff hadn't prepared for the possibility that the Forest Service chief would increase the level of protection for the peninsula's endangered spotted owl population. Many sales already in the pipeline didn't meet the new standards, and another two years would be required to prepare a new set of sales meeting the guidelines. In the timber towns of Forks and Aberdeen, committees sprang up to push for timber sales at the previous level. They didn't address the export issue. "The export policy has been disastrous," said a Hoquiam logger. "But we aren't going to talk about it. For once this community is united. We want a change in the Forest Service allowable cut. If we started talking about exports, it would tear this community apart."

The independent mills--those that didn't have their own timberlands--were at the mercy of the public land managers and the big timber companies. The premium prices offered for timber by Asian buyers priced the independent mills out of the market for timber grown on industrial or state land. National forests became virtually the sole supplier for the independents--and that supply was about to drop. The cutback in federal timber sales came at the worst possible time. The supply of timber from the industrial lands was about to drop, with the prospect of even higher stumpage prices.

The big timber companies, too, were going to have to learn to live with less timber. Much of the old-growth logging on the industrial lands was so recent that the second growth generally was far too young to cut. Nearly two-thirds of forest industry land in western Oregon held timber under thirty years of age in 1988. The most abundant age class was ten to nineteen years old. Since Douglas fir sawtimber normally isn't harvested on private lands until forty to sixty years of age, industry faced a big gap between its rotation age and the age of its standing timber. That gap translated into a reduced private timber supply.

Industry simply couldn't grow enough wood on its own land to offset the depletion of the old growth. In time, there was no reason that ample timber supplies couldn't be produced on twenty-two million acres of commercial forestland in western Washington and western Oregon. But for at least the coming two decades, industry was facing a significant drop in its timber production.

Economists debated just how severe the effects of "the gap" would be. A rather optimistic projection made in 1982 by economists in the Washington Department of Natural Resources showed less than a 4-percent drop in the timber harvest on industrial lands west of the Cacades between the 1970s and the decade after 2000. That projection assumed current management practices. If timberland owners were to intensify their management by more aggressive replanting, brush control, and thinning, the timber supply would increase by 4 percent. Those projections soon came to appear unrealistic, as intense logging of both old growth and second growth further depleted timber inventories. Some industry sources expected the inventory to be drawn down by another 25 percent during the 1990s.

When Oregon's Department of Forestry examined the timber supply in 1988, it forecast a drop of about one-third in harvests on industrial forests. That drop, expected around 1995, would never be recouped. The Oregon state projections closely paralleled figures put out by U.S. Forest Service economists in 1982. They predicted a sharp drop in the private timber supply, but recovery after two decades. The Forest Service projected a 44-percent drop in the timber cut on industrial lands in the Douglas fir region for the period 1976 to 2030.

"The gap" was largely due to past logging practices. The highly productive low-elevation old-growth forests cut before the 1940s almost invariably were left to regenerate themselves. Some unmanaged clearcuts have grown back as valuable forests that are being harvested today. Regrowth of others was stunted by poor seeding or competition from plants of little commercial value. "Had there been more foresight in the twenties and thirties," says Summit Timber Vice President Jack Dickson, "we would have so much timber available now it wouldn't be believable."

In the face of a reduced timber supply, capital flowed out of the Northwest to the Southeast, where the timber supply appeared more stable for the coming decades. Large corporations like Weyerhaeuser and Georgia-Pacific took capital accumulated from Northwest timber sales and made heavy investments in the Southeast. Some economists attributed this capital flight in large part to government interference in the market.

Chief among economists arguing for a more businesslike Forest Service is Barney Dowdle. Dowdle is a solidly built, ruddy-faced man with hands the size of Douglas fir logs. A professor in the University of Washington's College of Forest Resources, he prefers to work at home rather than at the college. His unbending, heretical views about the national forests have put him at odds with his colleagues.

Dowdle maintains that if the Forest Service would sell its timber in response to market demand rather than in accordance with its current version of sustained-yield forestry, timber companies would be more inclined to invest in the Northwest. "The state of Oregon has a commercial softwood timber inventory twenty-five percent larger than the entire South, including the states ranging from Texas all the way around to Virginia," Dowdle observes. "Oregon has more timber than that whole region, and yet we see industry leaving Oregon and going down there. Washington's commercial timber inventory is only about ten percent less than the entire South. If you take Washington, Oregon, and Idaho, you have about half the nation's softwood timber inventory and that's the raw materials base for the softwood lumber and plywood industries. The gross statistics ought to lead someone to the conclusion that *something* is wrong."

That something, Dowdle believes, is the Forest Service's policy of selling timber on a "nondeclining even-flow" basis. If the Forest Service were run like a corporation, it would liquidate its last old-growth stands much faster in order to bring greater revenues to the agency's shareholders, the taxpaying public. Because there is little, if any, net growth of merchantable timber in old-growth forests, it's more profitable to liquidate those forests quickly than slowly. That fact has not been lost on the industrial landowners. Receiving cash for their old-growth timber, they were able to reinvest the cash in other profitable ventures. They also were able to plant a new forest that would generate future revenues. The present net value of an old-growth stand was greatest if it was cut immediately, least if it was left on the stump indefinitely.

The big timber companies' accountants weren't unduly disturbed by the fact that there might be years with no timber sales because of the resulting "gap" in the age classes of the second-growth forests. If that gap meant that some timber-dependent communities became ghost towns, that wasn't the concern of corporate headquarters.

Dowdle points out that the Forest Service doesn't account for carrying costs of its old-growth inventory, disguises the capital costs of roads and reforestation as operating costs, and routinely sells timber at a loss. "It isn't even good socialism," he fumes. Other classical economists share his view. Marion Clawson in 1976 put the value of the Forest Service's "excess" inventory of mature and old-growth forests at $12 billion. At modest interest rates, Clawson calculated that the Treasury was losing $600 million a year-- the hidden costs of carrying that capital.

"One can readily imagine the reaction if every citizen of the United States was asked to contribute $3 annually toward the maintenance of an excess inventory of old trees that he might never see," Clawson observed. It's not at all clear, though, what citizens *would* say about that cost. Would they be any less inclined to forego revenues from old-growth timber sales than they are to forego revenues from a dam that would flood the Grand Canyon? A clue is offered in a survey commissioned by Rep. Al Swift of Washington when he was considering running for a Senate seat in 1987. Voters were asked if they would agree with a senator who "supports keeping small timber companies operating, even if it means clear cutting some virgin forests that are located outside wilderness areas." Thirty-six percent of the voters said they would agree with clearcutting, while 53 percent would disagree. The Forest Service's even-flow policy is about the only thing that has saved old growth outside the high-elevation wilderness areas. And if Swift's survey accurately reflects the sentiments of voters in a timber state, they want to save the forests.

On at least one point, ordinary citizens agreed with the laissez-faire economists. As taxpayers, they were growing tired of losing money on timber sales and in some cases losing valuable public timber to theft and fraud. Theft from the public lands came as something of a shock to many citizens. After all, hadn't the forest reserves been created a century earlier to stop just this kind of activity? Yet a series of criminal cases in the late 1980s suggested that mismanagement of the national forests remained widespread.

Federal prosecutors obtained convictions of three firms operating in the Olympic National Forest, including one of the largest independent mills on Puget Sound, for conspiring to rig bids. In a separate case, four loggers pleaded guilty to charges that they had stolen timber worth hundreds of thousands of dollars from the Olympic Forest's Quinault Ranger District. A Forest Service timber sale officer in the same ranger district was jailed after admitting that he had secretly subcontracted to do reforestation for the very loggers he was supposed to be supervising. A timber sale officer working in the Willamette National Forest, was jailed after he allowed a logger to cart off 244 truckloads of stolen timber and then accepted two truckloads of lumber from the thief.

In the absence of a tough commitment to law enforcement, bid rigging and theft seemed to be running rampant. Illegality is still winked at in much of timber country. "Sure, what he did was wrong, but everyone had been doing it for years and years. He just got caught," a friend of an indicted bid rigger in the Olympic Peninsula town of Forks told a reporter.

The mayor of Darrington, Washington, a retired logger, told me he doesn't view it as collusion if two mills decide not to bid against one another at auction; it's just common sense. Poaching cedar logs from the national forests is just part of the underground economy. A young Darrington man, discussing variable tree growth rates, described to me an 800-year-old four-foot-diameter cedar log that he and a friend had heisted from the national forest.

Even worse for the Treasury, the Forest Service actually manages to lose money on much of the old-growth timber it does sell. The agency's largest expenses, by far, support timber sales. Only in the Douglas fir region do timber sales typically produce a positive cash flow for the Treasury. By any measure, most federal timber sales lose money. It would be difficult to interpret below-cost sales as anything other than a subsidy to the timber industry. Nowhere have they become such a national scandal as in the Alaskan panhandle.

The rain forests of Southeast Alaska support the largest populations of grizzly bears and bald eagles in North America. Blessed with 200 inches of annual rainfall, the Sitka spruce and western hemlock stands of the Tongass National Forest represent the northernmost reach of the coniferous forest system of the West Coast. It is the least-fragmented coastal forest remaining in North America, but federal timber sales are rapidly changing that. What sets the Tongass timber sales apart from others--aside from the heavy environmental cost--is the extent of the federal subsidy. For every taxpayer dollar spent to support timber sales during the 1980s, timber buyers paid as little as ten cents. Two unusual arrangements drive the grossly below-cost Alaskan timber sales. One is the fifty-year contracts under which the Forest Service guarantees ample supplies of low-cost timber to two large pulp mills. The other, resulting from a political deal cut in the Alaska Lands Act of 1980, gives the Forest Service a permanent--and unprecedented--appropriation to support annual timber sales of 450 million board feet. That's more timber than the market can support. During the mid-1980s, environmentalists mobilized broad political support for legislation to cancel the permanent timber appropriation to the Tongass and to renegotiate the sweetheart deals with the pulp mills. Before its defeat in the Senate in 1988, the Tongass Reform Act received overwhelming support in the House of Representatives and appeared headed to passage in a future session of Congress.

Even in the Douglas fir region, the Forest Service routinely sells timber at a loss to the Treasury. Those losses aren't apparent on Forest Service balance sheets, however. Money-losing timber sales are disguised by the revenues of profitable sales within a national forest. To see the full story, it's necessary to look closely at individual timber sales. The most painstaking and revealing analyses of Forest Service timber sales are those done by Randal O'Toole. O'Toole, a forest economist, is founder of the Eugene-based Cascade Holistic Economic Consultants (CHEC). When O'Toole examined timber sales made by the Olympic National Forest between October 1985 and March 1987, he found that the U.S. Treasury lost money on one-sixth of the sales. On one below-cost sale, the high bid was $368,000. The purchaser received $254,000 credit for road construction and paid $90,000 into the reforestation trust fund. That left $24,000 for the Treasury. But because the Treasury paid the usual 25 percent of the gross sale price to the county government, taxpayers lost $68,000 on the deal. That's not all. The government also had to pay for additional road design and construction and the costs of preparing the sale for bid. When these costs were figured in, the total damage to the Treasury was $335,000.

"In sum," O'Toole concluded,

> despite the fact that the Olympic includes some of the most productive timber land in the National Forest System, forest managers apparently do not hesitate to lose money on sales. Depending on the definition of below-cost sales, from one-fourth to three-fourths of the timber sold by the Olympic since October 1985 was sold below-cost.

Below-cost sales frequently are paired with profitable sales in a process known as "cross-subsidization." On some rugged mountainsides in the Cascades and Olympics, timber values are too marginal and logging costs too high for timber to be removed profitably. That doesn't necessarily stop the Forest Service from selling the timber, however. Timber sales typically are offered in scattered clearcut units within a watershed. A timber buyer may be forced to take a loss on some of the units--"punishment units" as they're sometimes called--in order to bid on the units that make economic sense.

Why does the Forest Service put punishment units up for sale? In some cases, the agency sees the sale as helping to expand its already-extensive road network. The sale might help it meet its timber sale targets. Unprofitable clearcuts aren't unheard of on industrial lands, either. Landowners sometimes take a loss in order to replace a stand of vine maple and alder, say, with more profitable Douglas fir. But the Forest Service's below-cost sales are more extensive than anything industry would contemplate.

O'Toole attributes this fiscal mischief to bureaucratic empire-building. Rejecting the usual interpretation of environmentalists, who perceive the Forest Service as being in the clutches of industry, O'Toole contends the agency does what all organizations do: it maximizes its budget. Every timber sale made by the Forest Service boosts the agency's budget because timber buyers are required to pay into the reforestation trust fund established under the Knutson-Vandenberg Act of 1930. The K-V trust fund is in addition to congressional appropriations, as are special funds for brush control and timber salvage sales.

Clearly, the Forest Service is not run in the manner of private enterprise. Profitability isn't the agency's primary goal. A far higher priority to the Forest Service is fostering economic stability in timber-dependent communities. The nondeclining even-flow policy is intended to promote community stability. By maintaining a constant flow of timber, the agency attempts to insulate small towns in timber country from the boom-and-bust cycle they would otherwise face. Some economists, like Dowdle, reject the very goal of stability as tantamount to economic stagnation. Others argue that nondeclining even flow doesn't even promote stability because it doesn't take into account what's happening on private lands. Robert H. Nelson takes this position:

> The public land agencies have justified the even-flow policy partially as a means of promoting community stability. However, this implicitly assumed that private and state timber harvests were also being maintained at an even flow. As will be recalled, private timber has generally been harvested ahead of federal timber--indeed, for good economic reasons. But private supplies in the Pacific Northwest are now rapidly being depleted and will decline sharply until second growth forests become available. In such circumstances, an even-flow policy for public forests perversely becomes a destabilizing rather than a stabilizing influence.

If the national forests are to promote a stable timber supply in a region, it's arguable that federal timber sales should be a mirror image of the industry's timber cuts. When private lands are depleted in the wake of liquidating old growth, federal sales would be accelerated. When the federal timber supply inevitably drops, industry would boost the harvest of its second growth.

This is how the Shelton Cooperative Sustained-Yield Unit works. Federal and private lands were combined into a single "working circle" in which the annual cut would be limited to the sustained-yield capacity. Combining the ownerships administratively didn't increase the working circle's ability to produce timber, but it increased the rate at which the old growth could be logged off the federal land. When the sustained-yield unit was established, Simpson Timber's land was largely cut over and the national forest land was mostly intact. Simpson, with its monopoly on the federal timber, quickly shifted its logging from its own land to the national forest. The sustained-yield capacity of Simpson's 158,000 acres was added to the Forest Service's

111,000 acres. This allowed Simpson to take as much wood off the 111,000 acres as the Forest Service normally would sell on a 269,000-acre block. The result was a rate of clearcutting perhaps seen nowhere else on the national forests.

Simpson and the towns of Shelton and McCleary have benefited. The timber company gained a larger, more stable timber supply. The sawmills that underly the towns' economies were able to keep operating during a period when dwindling timber supplies otherwise would have likely shut at least one mill. Jobs in those mills became more secure than most jobs in sawmills. The Forest Service found a way to meet its longstanding goal of promoting community stability.

There is no free lunch, however. Simpson's monopoly on federal timber has done nothing to help scores of other mills at a time when the supply of timber has been falling. The independent mills of nearby Grays Harbor recognized that danger in the 1940s when they quashed a sustained-yield unit the Forest Service tried to set up there. The lack of competitive bidding in the sustained-yield unit has reduced federal revenues from timber sales. The logging blitzkrieg on the Shelton Ranger District is widely viewed as evidence that cooperative agreements are a recipe for environmental destruction.

Appalling though the Shelton landscape is today, the Quinault Ranger District to the west or the Hoodsport District to the north may not look so very different in a few years. The ancient forests are being stripped off all the ranger districts. It's just happened faster in Shelton. This is not to suggest that environmental damage from the breakneck clearcutting of a broad landscape isn't especially severe. It is. But the viability of the old-growth ecosystem depends more on today's decisions over which forest stands will be cut than how fast they will be cut.

Putting aside the excesses of the Shelton arrangement, it offers a useful lesson in sustained-yield forestry. Combining public and private lands into a sustained-yield unit *can* boost timber production long enough to bridge "the gap" in the timber supply. That boost could be a useful tool in settling the war over the ancient forests.

Environmentalists won't sit still for Shelton-style overcutting on other ranger districts. Nor will independent sawmills allow another Shelton-style monopoly. It's possible to envision any number of scenarios in which the national forests' sustained-yield base could be increased without making the mistakes that were made in Shelton. If such an approach were to work, independent sawmills would have to be included in on the deal, and environmentalists would have to win additional protection for the old growth. "We could do that," said regional forester Jim Torrence shortly before his retirement from the Forest Service. "You give me control of the private lands. I question whether those private lands will ever reach maturity. I think it's being loaded on boats at an inordinate rate and going overseas."

Twenty-one years after an old-growth forest was clearcut (above) on Weyerhaeuser's St. Helens tree farm near Longview, Washington, a new forest has grown up (below).

Top: *The forests of the Olympic National Forest's Shelton Ranger District have been cut rapidly under an unusual agreement between the U.S. Forest Service and Simpson Timber Company. Bottom: When a new forest grows there, it will look less like the forest that preceded it than like this sterile second-growth stand on the nearby Quilcene Ranger District.*

A century after it occurred elsewhere in the United States, the timber industry in the Pacific Northwest is still making the transition from an old-growth base to second growth. The 1980s represented the turning point when the bulk of the region's timber supply switched from old growth to second growth. Although old growth would continue to account for most national forest timber sales well into the next century, a major transition was being completed on the private lands.

In western Oregon, old growth remained on just over one percent of private timberlands in 1987. Virgin forests accounted for one-fourth of the timber cut on private land in Washington as late as 1984. The few timber companies that had old growth left--most notably Weyerhaeuser and Burlington Northern's Plum Creek Timber Company--were cutting their last big trees as rapidly as possible. The forests would be gone by the mid-1990s, possibly leading to a crash in the local timber economies of some communities.

Some observers, such as Weyerhaeuser economist Dick Pierson, chose to look on the bright side of things. The Pacific Northwest, he said, would be "the first country, the first region in the United States, to success-fully transition between an indigenous old-growth natural forest situation to a managed second-growth situation without prolonged and severe impacts on harvest levels."

Indeed, by historic standards, the transition was going well. In other parts of the country, the timber industry had simply moved on to other states after the forests were stripped bare. The Northwest learned something from those painful lessons. Reforestation became the industry standard during the 1940s. The region's economy, Washington's more than Oregon's, diversified with a host of new high-tech industries. The timber industry made itself more competitive through automation and tough dealing with labor unions. After weathering the market slump of the early 1980s, the industry's biggest concern was the timber supply.

The irony remained that sustained yield sometimes was given lip service without action. A multinational corporation could say it was practicing sustained yield when it shut down operations in the Northwest and acceler-ated operations in Canada, the Southeast, or the tropics. Forest depletion hasn't ended on industry lands. Bob Dick, the Washington Forest Protection Association's director of forest management before becoming Alaska's state forester, is concerned about landowners "robbing the cradle" by cutting timber prematurely. "We had a period of very difficult markets in the mid-eighties," he explains. "We had a lot of landowners who were cutting very heavily during the eighties just to keep the doors open. There were periods in the late Eighties when the reverse was true; we've had good markets and people have been cutting heavily to get well."

Gus Kuehne, president of Northwest Independent Forest Manufacturers, similarly, lambastes large corporations for the "greed" that leads them to make a quick buck by selling immature timber on the export market. "For example," he says, "I don't think twenty years ago anyone would have believed that ITT Rayonier would be cutting over all that second growth out of Forks that they are today. But they discovered a China market and somebody wanted to capitalize on that revenue right away, so they cut it all quicker than I think anybody ever anticipated they would, including them."

The impending drop in the Northwest timber supply doesn't just mean lost jobs and a gap in industry revenues. It also creates pressure for the liquidation of old-growth forests. There are other ways of dealing with "the gap." Something could be done about the "sealift" that was carrying logs to Asia by the billions of board feet. Production could be increased on nonindustrial private forests. Incentives could be adopted to stop the conversion of the best private timberlands to urban sprawl. Some national forest lands could be combined with private lands for the purpose of calculating sustained yield.

Any of these actions would boost the amount of timber available to Northwest sawmills. Instead, the forest-products industry set its sights on maximizing the federal sales of old-growth timber. Ancient forests managed by the Forest Service and the Bureau of Land Management offered the industry's cheapest, most convenient timber source. The old-growth forests were being logged by default.

The "Tarheel fence"--a log laid parallel to the dirt road--didn't work. The four-wheelers went right over the log and made mincemeat out of Dave Larson's stand of young Douglas fir. Bob Hoey, Larson's tree farm manager, explains that he arrived on the scene too late to prevent the damage. The joyriders had brought two four-wheelers in a trailer towed by their El Camino. They careened around three or four acres, riding right over the Christmas-tree-shaped firs that stood eight to ten feet high. Hoey later staked thirty or forty of the supple trees to straighten them up. Larson shakes his head. "I would never have dreamed that people would deliberately run over six-year-old trees."

The damage isn't severe. It's just galling. Trespass isn't unusual out here in the woods near Darrington. During the eleven years that Larson has been growing and harvesting trees near the confluence of the Sauk and Suiattle rivers, he's had his share of encounters with drunken trespassers wielding guns. One even ran him off his own property. Not that Larson minds hunters who are reasonably polite. They keep down the number of bears and deer that sometimes go after his crop.

Dave Larson is a tree farmer. There are many independent tree farmers in Oregon and Washington. There aren't many, however, who, like Larson, depend entirely on this business for their livelihood. But he's an independent sort, which helps. The earth rumbles when this basso profundo speaks with the impossibly deep voice that has won him recent roles in the Everett Opera's productions of *The Magic Flute* and *The Bartered Bride*. He explains that he came to tree farming somewhat circuitously. Formerly a CPA, Larson gradually built his assets by investing in real estate partnerships with his clients. He managed or participated in apartment projects, office buildings, and recreational developments ("We screwed up some good pieces of land"). Each time a project was completed, Larson had to reposition his capital. He searched for a longer-term enterprise in which he could make the land work for him "without selling the farm."

Larson grew tired of the pressures of accounting and real estate development. "I finally got to looking around western Washington and observed what I should have seen in the first place, and that's the trees." Studying up on forestry and the timber market--which was heating up in the mid-1970s--he decided this was the business and lifestyle he was looking for. It didn't take long for Larson to figure out how much land he would need to bring in the kind of income that would support him and his family in the comfortable style to which they were accustomed. With 2,400 acres of timberland, his family could harvest an average of forty acres annually on a sixty-year rotation.

The trick would be twofold: to buy the land at a reasonable price (the presently merchantable timber should be worth at least two-thirds as much as the total cost of the property) and to maintain adequate cash reserves. The reserves would be critical so he could sell his stumpage when timber commands a good price. "When the market is down, that's when some tree farmers, in order to make payments on purchase contracts or send children to college, are forced to liquidate. That's too bad. It doesn't help that specific tract and it doesn't help the long-term health of the forest-products economy if a timber stand is harvested prematurely."

Buying timberland in western Washington proved more difficult than Larson anticipated. He found that industrial landowners like Weyerhaeuser, Scott Paper, and Georgia-Pacific owned most of the good tree-growing properties of any size. When the major sellers put land on the market, much of it was bought by corporate investors from Germany and other countries. Larson eventually assembled his acreage by buying scattered tracts, typically 80 to 160 acres, from various sellers. The first acquisition included parts of two old homesteads on the Sauk and Suiattle rivers. "What people like myself are finding," he says, "is that what that country does best is grow trees. Wonderful tree-growing country, but it must have been a miserably wet homestead."

Independent landowners, most of them with smaller holdings than Larson, control a significant amount of the available timber in the Douglas fir region. These "nonindustrial private forests" account for more than one-fifth of the commercial timberland in western Washington, one-sixth in western Oregon. But the timber-producing potential of these lands isn't being realized. The numbers show this: in western Washington, where non-industrial forests make up 22 percent of the timber land base, these smaller woodlots hold only 14 percent of the timber inventory and account for only 10 percent of the harvest.

This "Third Forest," as the nonindustrial private forests are sometimes called, clearly isn't producing the timber it could. (The first and second forests are on land owned by industry and by state and federal government.) The independent mill owners, though, tend to focus on timber in the national forests. This isn't because these forests are the best tree-growing areas but because they contain the last large blocks of old growth and because that timber is relatively cheap. It's also convenient. Sawmills find it far easier to buy five million board feet of timber at a Forest Service auction than to haggle with Dave Larson and half a dozen other owners over the trees on their scattered small tracts. Larson currently doesn't sell timber from his Sauk-Suiattle tree farm to nearby Summit Timber because he won't accept the low prices that Summit now pays for federal timber. Why should Summit, for its part, pay Larson the amount that a Japanese trading company or Weyerhaeuser pay when the company can buy wood more cheaply from the Forest Service?

Private forest owners like Dave Larson are sitting on an important timber reserve. While timber inventories were declining on industry lands and in the national forests, the volume on these smaller woodlots in Washington grew by nearly half a billion board feet during the 1980s. Owned by everyone from the retired Boeing worker with a twenty-acre vacation lot to investment groups managing thousands of acres for timber production, these lands are producing nowhere near the amount of timber they could. Much of the nonindustrial land isn't stocked with the softwood species that are the backbone of the Northwest timber industry. More importantly, 44 percent of Oregon's Third Forest has been so neglected that it is classified as unmanageable. Management is no better in Washington.

The saddest aspect of the failure of these private lands to produce is that they include some of the most productive timberland in the nation. Charlie Krebs, director of cooperative forestry programs in the Forest Service's State and Private Forestry division for the Northwest, says of the nonindustrial forests: "They tend to be down in the valleys. . . . They're not up on the rocks and ice where the national forests are. They can grow timber very well. We're looking at very high site classes. What it's doing is growing brush very well right now."

Statistics bear out the unrealized potential of the nonindustrial forests. In western Oregon, researchers at Oregon State University found more acres of "high" and "medium" site classes in the Third Forest than on national forest land. If the nonindustrial lands were managed more intensively, concluded the study headed by John H. Beuter, timber harvests could be increased three- to fourfold. That conclusion was said to reflect "a working capability, not an unrealistic optimum based upon all lands producing timber at full potential." In western Washington, similarly, the Department of Natural

Resources reported that an impressive 71,000 acres of nonindustrial forest was in the highly prized Site Class 1. In the national forests' commercial land base of equal size, there were only 16,000 Site Class 1 acres.

One of the obstacles to fuller production on the nonindustrial lands is the lack of a strong public commitment to promoting forestry. Many forest owners have little or no interest in commercial timber production. Some prefer to replace their trees with pasture or scenic views; others want a natural, unmanaged forest with the richest possible wildlife habitat. Too few small landowners realize there are ways of managing at least part of their land for these goals while growing a commercial timber crop. Without professional advice, many decide forestry isn't worth the trouble or the risk.

Bill Arthur, Northwest representative of the Sierra Club, recalls his trouble trying to sell timber from a portion of his 42-acre property on the Pend Oreille River. He had to contact four logging companies before finding one that would agree to the selective cutting and careful yarding of logs he insisted on. "I had two of them just flat-out tell me that if we weren't going to clearcut a block of at least ten to twenty acres they weren't interested in it," Arthur remembers. "If I was your average landowner and went to two timber firms and told them what I was interested in doing and was told they couldn't do that, I probably would have quit looking. So you wonder how many people want to use their land intermittently for fiber flow as long as it's consistent with whatever values or long-term objectives they have for their piece of land."

Marion Clawson contends that most merchantable timber on smaller woodlots will eventually be sold--by a future owner if not the present one. But future owners won't have much timber to sell if present owners fail to plant commercially valuable species and follow up with brush control and thinning. The federal government, in cooperation with the states, offers advice and cost-sharing to small forest owners. The Reagan administration, favoring money-losing timber sales of the publicly owned old growth, gutted programs to boost production in the fertile Third Forest. The budget for cost-sharing and technical advice was slashed from $19.4 million in 1981 to $9.7 million in 1986--slashed rather than killed because Congress rebuffed the administration's attempt to "zero out" the program. The Bush administration sought additional, deep cuts. The savaging of service forestry represents a sharp about-face from the emphasis that previous administrations put on the Forest Service's State and Private Forestry branch. Nixon's Advisory Panel on Timber in 1973 called nonindustrial forests "the listless giant of forestry" and urged stronger efforts to boost timber production on those lands.

The unrealized productive capability of second-growth timberlands isn't restricted to the nonindustrial side. Industrial and nonindustrial forests alike are being taken out of production at an alarming rate. Washington's Department of Natural Resources (DNR) estimates that 235,000 acres of private forestlands in western Washington will be converted to roads, utility corridors, and residential and commercial development over a two-decade period. Between 1980 and 2030, DNR projects a loss of 523,000 acres, or 5 percent of the timber base in western Washington. Given the high quality of this timberland, these losses could have a greater impact on the timber supply than the Oregon and Washington wilderness acts of 1984 have had. The greatest loss of forests is occurring in the fast-growing urban areas of Puget Sound and the Willamette Valley. Serious tree farmers find it hard to resist the temptation to convert timberland to housing tracts. The profits of conversion are substantial, and the penalties for holding the line can be high. The exurban commuters who move next door to tree farms complain when trees are cut or when herbicides are sprayed. They trespass with their motorcycles and three-wheelers, damaging young trees and creating a liability problem. Tree farmers are assessed for local improvement districts that bring sewers and roads--and more problems--to their forests.

Instead of working with local government to find ways of controlling urban sprawl, some timber companies are choosing to hasten the demise of forestry on the urban fringe. Weyerhaeuser drew brickbats in the late 1980s for its plan to retire 2,000 acres of rich second-growth forest and build a combination residential development/business park/golf course east of Seattle. Three thousand six hundred homes would be built in the Snoqualmie Ridge planned community not far from the timber company's Snoqualmie Falls sawmill. The proposal didn't mean that Weyerhaeuser was about to turn its back on the forest-products business. "As long as George Weyerhaeuser is in charge, Weyerhaeuser will be a timber company," says one mid-level executive. But some Weyerhaeuser-watchers saw a shifting corporate emphasis in the promotion of John W. Creighton, Sr., from chief of the company's real estate development division to the presidency of the corporation in 1988, while the land-use battle was raging over Snoqualmie Ridge. George Weyerhaeuser moved up to chairman of the board.

The withdrawal of such fertile timberland disturbed others in the timber industry. While large companies like Weyerhaeuser could increase profits by taking assets out of timber production, the independent mills were suffering from a declining log supply. But the independent mill owners were reluctant to challenge Weyerhaeuser's right to do what it wants with its land. They didn't want anyone telling them what they could do with their property. Besides, they already were fighting over what at the moment seemed like a bigger prize: log exports. Steve Wells, a forester in King County's Planning Division, is upset over the silence of the small mill owners and their industry association, Gus Kuehne's Northwest Independent Forest Manufacturers. "There are big chunks of our forestland base being converted," Wells says. "I'm not saying they shouldn't be. But when you get Gus Kuehne and his small-mill people screaming about national forest withdrawals, you don't see Gus Kuehne screaming about forestland conversion on the Snoqualmie plateau. They're rolling over and playing dead about the conversion of low-elevation, rather high-quality site lands."

During Dave Larson's decade in the business, he has come to call himself a "missionary" for tree farming. Having given up the fast bucks to be made from real estate deals of dubious merit, he's developed an attachment to the land and to the forest-products business.

He's parted company with the prevailing financial wisdom that equates good business practices with converting physical assets to cash. To him, it makes good business sense to watch his Douglas fir trees grow toward a harvest that won't take place for many years--in some cases after he expects to be around to sell them. Committed to long-term productivity and sustained yields, he's passed up some lucrative opportunities to sell cedar at high prices. His forester friends kid him that he's secretly given names to each tree. Let them laugh, he responds. "My conclusion is that in timberland, the investment play is in harvesting only as much as you need to pay bills and in letting the rest continue to grow."

Larson is no Sierra Club activist. He gets upset over what he perceives as the excesses of "the preservationists." Unlike his wife, who enjoys backpacking in the wilderness, he prefers to stroll among the growing trees on his own land. He reacts with some amusement at the eagerness with which his tree farm manager, Bob Hoey, offers his opinion about ancient forests during my visit to the Darrington tree farm. Hoey, the son of a gyppo logger, points to the old-growth clearcutting on nearby Prairie Mountain and calls it "ecocide." He's reluctant to go with a friend to earn good money logging old growth along the Olympic Peninsula's Humptulips River. "I don't like cutting down the forest," he says. "I don't think they should be logging any more old growth."

Larson is more concerned about the management of second-growth forests, and he's disturbed by some of what he sees. As we drive through the Stillaguamish River valley between Darrington and his home on the Tulalip Indian Reservation, we see many hillside forests in which commercially valuable conifers are being choked out by straggly deciduous trees. "You can drive on Highway 9 from here to Woodinville," Larson frets, "and see mile after mile of land that's been allowed to grow up in brush. It breaks my heart. You can see that it's not being held by active tree farmers. It's being held by land speculators."

I'm still thinking about his words as we pull into Larson's driveway. Land speculation is part of our dream of never-ending growth and "progress." Accommodation must be made for Puget Sound's rapid growth. But dubious land-management policies down here in the lowlands are only exacerbating the threat to the old-growth forests up on the hillsides. Everyone seems to want a little piece of land in the country--the "country" quickly becoming a sprawl of one- to five-acre tracts cut up by commuter routes. The cost for this is that the low-elevation forests are being destroyed in the process. Some of the nation's most productive timberland lies untended while local sawmills are closing down for lack of logs. A parade of trucks continues to carry raw logs to the docks. And then the irreplaceable old growth is auctioned off to forestall a timber shortage that has parallels to the "timber famine" that never came.

Surely there's a better way. Surely the private lands can supply the mills if the mills are important to us. Eighteen million acres of the world's richest second-growth forest lies on the west side of the Cascades. Up above, in the rugged hills and valleys, lie four million acres of land less productive for growing trees but covered by a spectacular and endangered ecosystem. Surely we don't have to sacrifice the majesty of the old growth just because we're squandering the bounty of the second growth.

7

Brave New Forest

From the road, the green on the fields could be mistaken for any low-lying midsummer crop. Sprinklers are pumping water over parts of this 300-acre farm. Without a closer look, the crop could be leaf lettuce, celery, or carrots. In fact, it's Douglas fir, along with smaller quantities of hemlock, noble fir, Sitka spruce, and grand fir. The property probably looks about the same as it did when the previous owners grew sweet corn and green beans.

It's Weyerhaeuser property now. Dick Pierson, the Weyerhaeuser economist at the wheel of this company van, points to a row of alders on an adjacent property as we drive onto the Mima tree nursery south of Olympia, Washington. "That's what we *don't* want to see as foresters," he says of the deciduous trees. Conifers are what commercial tree farming in the Pacific Northwest is all about. Not just any conifers. Only the fastest-growing, hardiest conifers that produce the best-selling wood products. That means Douglas fir wherever growing conditions favor this king of the timber industry. No one surpasses Weyerhaeuser in the business and technology of growing Douglas fir.

The huge white shape of Mount Rainier provides the backdrop as we walk into a huge field of "Two-oh" firs. These seedlings, sown by machine, spend two full seasons in the ground before being pulled up, boxed in the packing room, and then stored in the cooler until time for replanting in a clearcut.

Other fields here are stocked with "Two-ones," seedlings that are transplanted after two years to another nursery bed, where they spend another year before packing. The larger two-ones are used to reforest sites where deer, rabbits and mountain beavers pose problems. Eighty to 90 percent of the seedlings typically survive in the field. Here in the nursery, the two-ohs are tightly packed, twenty-five to the square foot. But they look extremely healthy. The soil here is a porous mix of loam and sand. "This is considered very droughty and of poor fertility, but you can always add water and fertilizer," says nursery manager Tom Stevens. A nursery in the Nisqually River valley had to be abandoned because it was too moist.

The first of Weyerhaeuser's eight nurseries, the Mima facility ships out twenty-five million seedlings a year. Half are used by Weyerhaeuser, the other

140

half sold to other tree farmers including state and federal government. Mass production of seedlings for reforestation began at Mima in 1967, as the cornerstone of Weyerhaeuser's high-yield forestry program. The roots of the program go back to the 1930s, when company foresters started experimenting with reforestation. "Timber is a crop," the company proclaimed in its advertising as early as 1937. The Clemons Tree Farm near the Olympic Peninsula town of Montesano was dedicated in 1941. By that time, it was obvious to all who looked more than a few years ahead that the timber industry's future depended on its success in growing new forests. The end of the old growth that had once seemed inexhaustible was in sight. Not all landowners were quick to jump on the reforestation bandwagon. But today, reforestation and the "intensive management" practices that go with it are standard operating procedure in the industrial forests.

In the early days, reforestation typically was accomplished by aerial seeding. The national forests laid out clearcuts in smaller patches to allow natural seeding by adjacent trees. With aerial seeding and natural revegetation alike, the results were less than spectacular. Even when the seeds sprouted and grew--which wasn't always the case--they had to compete with other plants that thrive in disturbed soil and bright sunlight. Browsing animals often made quick work of young conifer shoots. Slowly, the idea grew that it made sense to plant seedlings by hand. Initial costs would be higher for this more labor-intensive process, but the benefits were obvious. Planting two-year-old trees instead of broadcasting seeds gives the new forest a two-year head start. That translates roughly into a four percent productivity gain, assuming a forty-five year rotation as Weyerhaeuser currently uses. Actual gains may be higher because planted seedlings are more likely to survive to maturity and because stocking levels are controlled. (If a forest is over-stocked, more money must be spent to thin it.)

Planting is just one part of intensive forest management. Before planting, it is common (but no longer universally practiced) to prepare the site by burning slash. Unhampered by a jungle of branches and other woody debris, reforestation crews can do their job more quickly and thus with less expense for the landowner. Slash burning reduces the amount of debris that can spread an unintended fire into adjacent stands, and it converts nitrogen into a form usable by the young trees about to be planted. After planting, competing vegetation is controlled either by manual brush clearing or by use of herbicides. On the large industrial forests, aerial herbicide application is the usual method. The damaging effects of some chemicals on animal life as well as plant life has led to controversy over their use. This is particularly true in areas where houses have been built next to commercial forests. Far fewer chemicals are used on forests than on food crops, because of the long rotation age of trees. Herbicides are applied only during the early stage of reforestation, before the shade of the growing forest's canopy eliminates any serious competition. While American farmers use herbicides annually or even more frequently, foresters might make two herbicides applications during a 40- to 100-year-rotation. Herbicides are applied annually to only 2.5 percent of private forests in western Oregon each year. On many stands, insecticides aren't used at all.

Two other activities complete the standard bag of tricks that comprise intensive management. One is thinning a forest stand, cutting down some trees to give their more vigorous neighbors room to grow. Thinning may be precommercial (when the thinned trees have no commercial value) or commercial. The other "trick" is fertilizing. Soils in the geologically young Pacific Northwest mountain ranges are often deficient in nitrogen and other organic materials. A little fertilizer, it turns out, can go a long way toward boosting timber yields. Soil scientist Dale Cole is among the researchers excited by the increased timber production that can be realized with fertilizer. At his office in the University of Washington's College of Forest Resources, he pulls out a slab of wood--the cross section of a tree trunk of modest size. The disk appears unremarkable until Cole directs my attention to the growth rings. Those near the center are very closely spaced, showing a slow growth rate. The outer rings are wide, less than two to the inch.

"This is from a naturally growing forest. And at this point, I intervened," Cole explains, indicating the spot where the rings suddenly widen. The way he intervened was to spray a load of municipal sewage sludge into the forest. "This site was about thirty-five years old when the sludge was put on it. It would probably never have reached merchantable size. It would have been hundreds of years before that forest could be looked at as large enough in size to be harvested. We took the disk eleven years ago and the growth rate has not decreased. You can go in right now and do the harvesting. So we've made it from an unproductive site to a productive site in a short period of time. The real point here is that you can improve the productivity of a site dramatically. The potential is incredible."

Metro, the Seattle-area sewage-treatment agency, was so impressed by Cole's demonstration that in 1988 it went out and bought 2,000 acres of forest near Yelm, Washington. Instead of continuing to lose money landfilling its sludge, Metro wanted to put it to work as fertilizer. Revenues from the resulting timber sales would give ratepayers some relief. Neighbors of Metro's new tree farm were unhappy, however, at the prospect of heavy metals and residual human pathogens being sprayed around their "backyards." In UW's experimental forest, shrews and their prey took up cadmium at high levels. Metro executives, facing a permit battle and public relations nightmare, considered dropping the whole idea.

Whatever comes of the sewage debate, Cole insists sludge will have its day. A different kind of sludge--the residue of pulp mills--is produced in even greater quantities than sewage sludge in the coastal forest region. Free of pathogens and heavy metals, this sludge may be a better bet for use as fertilizer. Waste materials probably will never supplant standard commercial fertilizers; there aren't enough of them. They represent just one more tool in a growing storehouse of materials and techniques used to grow trees bigger, faster, and stronger. Forest management is most intensive on the industrially owned forests. It is least intensive on the nonindustrial private forests. "It's just a crying shame," Cole says of the nonindustrial owners' failure to make their fertile lowland forests more productive. "So it just waits there for development to take place and of course we convert it to other kinds of uses."

The forest-products industry's technological leaders place a great deal of faith in their ability to continue increasing timber output through intensive management. According to Weyerhaeuser figures, intensive management-- planting, thinning and fertilizing--will increase the volume of timber produced by 170 percent in much of the Douglas fir region. During the past two decades, Weyerhaeuser has planted two billion trees on half of its six million acres in the West and South. Production-line techniques have come to the nurseries and tree farms. Yet this intensive management is just the beginning. To build the tree farm of tomorrow, scientists are using genetic selection techniques and cloning. In the not-too-distant future, the biotechnologists may even turn to manipulation of the trees' genetic structure. These approaches bring risk as well as promise.

Tree farmers joke about the way cones have been bought and sold in the seed trade.

"What elevation do these cones come from?" asks the buyer.

"What elevation do you want?" answers the seller.

There's more than a kernel of truth in the old parody of the cone trade. In the early days of tree farming, no one paid much attention to the source of seeds, as long as they were Douglas fir. Slowly the realization spread that Douglas fir seed from a low-elevation site in Medford, Oregon, might not be the best source for a mid-elevation tree farm in Bellingham, Washington. Buyers became more choosy, and sellers responded by setting up certification programs aimed at keeping track of the region and elevation from which cones came. But the basic process of collecting and marketing cones hasn't changed much. Individuals paid by the bushel still plunge into the forest to collect the cones they will sell to the merchants. And although certification programs have taken some of the uncertainty out of seed sources, much is still left to the integrity of free-agent cone gatherers.

Among the best-known cone sources are the woods around Darrington, Washington. The Douglas firs of Darrington probably don't grow any faster or produce better wood than the trees of nearby Sedro Woolley or the Oregon towns of Molalla or Sweet Home. But the old-growth firs of Darrington are better known because they happened to be the source of seeds that went to Europe in the early years of this century. Impressed with the growth of the trees from those seeds, buyers from Belgium, Italy, France, Austria, and

Germany still set up buying stations in Darrington. No one today even knows which part of the forest--much less which individual trees--those original cones came from. Occasionally, visitors from Europe and Japan want to see the seed trees of Darrington.

Jim Merritt, a retired Forest Service officer who works periodically as a field inspector in the cone business, recalls the time he took a foreign visitor in 1985 to Gravel Creek to see "this fantastic stand" of old-growth fir on the bottom land of the Sauk River drainage: "It was gone. There was nothing left. As far as I could see, it was cut over. All of a sudden I began to realize there's damn little large old growth--I'm talking about *large* old growth--left." First on the private lands, then on the public, the big trees of the Sauk valley were lost--and with them much of the area's genetic reservoir. During his days as assistant ranger in the Darrington District, Merritt recalls, "You took the big stuff first. Anytime you opened up an area, a road from the main Sauk road up into the other drainages, there were no deficit sales. You picked out the best stuff. More and more there's not much of that best stuff left. Now they're logging things we wouldn't even have looked at." The original trees of those rich bottom land forests are gone.

It doesn't really matter to a tree grower whether the genetic makeup of a tree would produce the greatest size after six, seven, or eight centuries. What's important is maximum growth during the crop rotation of his tree farm. The rotation for sawtimber in the Douglas fir region may be anywhere from 40 years on private land to 120 on parts of the national forests. Though the demise of the biggest trees may not upset a tree farmer, the narrowing of the region's genetic diversity is of concern to at least some in the industry. The industry's response hasn't been to leave diversity up to nature. Rather, it's been to take control of a diminished genetic base and cultivate a relatively small number of seed trees. In the case of Weyerhaeuser, no more than 3,000 trees are used as parent stock for reforestation on the company's three million acres in western Oregon and western Washington.

Dave Hodgin, a casually dressed young scientist with a serious demeanor, manages Weyerhaeuser's seed orchard in Rochester, Washington. The orchard is part of the company's effort to "genetically improve" its trees by selecting for maximum growth potential. "Although breeding to improve crops from dogs to cats to cattle to wheat has been going on for years and years and years, few people thought of timber as an agricultural crop--although it really is at this point," Hodgin explains. Seed from this "first-generation" orchard and similar facilities are the source of more than nine-tenths of Weyerhaeuser's Douglas fir seedlings in the area served by this orchard. By the early 1990s, all of the company's nursery growth will be genetically improved stock.

Genetic selection began in the mid-1960s, when scientists like Hodgin went into natural forests to find promising parent stock. "What we tried to select was a tree that was utilizing its square feet of ground the best." They

Left: The commercial forest of to-morrow may begin in the laboratory, where industrial-scale methods are being developed to clone trees. Bottom: The young genetically selected Douglas fir that cover Weyerhaeuser's Mima tree nursery will be transplanted into clearcuts.

looked for the tallest, broadest trees. They rejected trees with forked tops or excessive branching. They checked the wood density. ("It does no good to grow trees that are too soft to utilize.") They used increment borers to determine the trees' growth rates. From that effort, the Rochester scientists chose 120 parent trees for their new orchard. Tissue from each tree was cloned twenty times and the clones grafted onto the stumps of other young trees. The early effort was plagued with "graft incompatibility"--a sort of tissue rejection that caused trees to weaken and die. With improved techniques, the success of grafting rose from 50 to 90 percent. Today, the orchard is filled with neat, widely spaced rows of hearty Douglas firs.

Six orchards in western Oregon and western Washington developed genetic material at the same time. Weyerhaeuser executives looked closely at costs and benefits. "We had to satisfy the same long-term economic basis criteria that a mill did," Hodgin recalls. Production of hemlock was dropped in all but one orchard. Noble fir was put on hold, and further work with high-elevation Douglas fir was stopped. Only for lower-elevation fir would seed cultivation clearly pay its way. The first benefit for the company would be the availability of an assured seed supply. The more important goal, however, was the productivity boost expected from genetic selection.

For fourteen years, Weyerhaeuser researchers have been monitoring the performance of the progeny of their parent trees on test plots. Some parent trees have been "rogued out" as their offspring have proved to be less productive or more susceptible to disease or insects. The winnowing-out process is expected to bring continuing productivity gains. Weyerhaeuser hasn't released data on the field testing of its first-generation improved seed. But one published estimate projects a 12- to 15-percent productivity gain per generation of genetic selection. Cloning is expected to bring further gains. Weyerhaeuser scientists Peter Farnum and Roger Timmis joined with J. Laurence Kulp, the company's vice president for research and development in a 1983 *Science* article that suggested two generations of breeding and "intense clonal selection" of Douglas fir "could improve yield nearly 100 percent over that of wild seed." In 1987, Weyerhaeuser took the next step in genetic selection when it opened a second-generation seed orchard in Medford, Oregon. There, the trees from a first-generation orchard are being selectively bred for another round of productivity gains.

Before moving to second-generation genetic selection, Weyerhaeuser took the conservative step of expanding its first-generation pool of genetic material. Concerned that the original 120 parent trees per orchard might be too small a genetic base, 180 more trees were added. Hodgin describes the company's genetic improvement program as broadening, not reducing, genetic diversity. "We haven't had tremendous genetic variation out there for two hundred years," he says. "The old growth that was here before the white man came was your maximum genetic variation. Once logging took place, they didn't think about replanting 150 years ago. Your genetic variation went down all along the way. That's one reason Weyerhaeuser was concerned. . . The gene base has been preserved in orchards. Once those old-growth stands were cut, the genetic variation was getting narrower and narrower. Any acre we plant will have more genetic variation than it would with natural seeding."

Parent trees in Weyerhaeuser's seed orchards are pollinated naturally. At Rochester, 3 to 8 percent of the offspring are "pollen-contaminated," that is they receive genetic material from random sources outside the orchard. That contamination will be eliminated, if scientists in Weyerhaeuser's Technology Center succeed in perfecting techniques for cloning Douglas fir. In the basement of a modern building in Federal Way, Washington, brightly lit rooms hold racks and racks of glass dishes. Each dish in this clone bank contains the genetic material of a different tree. One hundred and fifty "genotypes," or genetically unique tissues, are stored here. It's an expensive process. Every few weeks, the tissue of each genotype must be chopped up and placed in a fresh growing medium. Not allowed to mature, the tissue is kept forever young. The "germplasm," as scientists call the tissue, is being stored at low temperature during the years that the performance of these genes is tested in the field. Initial field tests have shown that Douglas fir clones grow as well as seedlings and that clones show the desired uniformity of growth.

In the botanical equivalent of the test-tube baby, scientists in Weyerhaeuser's tissue-culture lab and elsewhere are trying to come up with economical, production-line techniques for creating genetically identical Douglas fir seedlings. They haven't yet succeeded. Dr. Cheryl Talbert, leader of Weyerhaeuser's effort to clone trees *en masse*, believes the most promising technology is the one known as "somatic embryogenesis." Beginning with a single cell, or a cluster of cells, the tissue develops like an embryo into a plantling that can be introduced into the field. Rooting of clones has been unreliable and some cloned trees planted in 1980 and 1981 have not grown straight. Talbert says, "That's the product of some pretty early techniques, and some of those look pretty weird. Some of those look great. There are a lot of bugs in the system." As tissue-culture techniques improve, researchers hope to develop a seedlike "bead" containing the somatic embryo, nutrients, and antimicrobial and antifungal agents.

If scientists come up with cost-effective cloning methods, they'll be able to stamp out as many copies as they want of an "elite" plant. Eventually, biotechnologists may use recombinant DNA techniques to design trees according to whatever specifications they wish. One scientist jokes about producing a rectangular tree. Another, who has inserted the gene that produces a firefly's glow into the molecular structure of a fir, chuckles that he may come up with a self-lighting Christmas tree. At Weyerhaeuser, where the corporate eye is planted firmly on the bottom line, there's no rush to splice genes. Rex McCullough, director of Weyerhaeuser's biological sciences division, sees no near-term gains for forestry in the "far-out technology" of recombinant DNA. For now, the company is investing its research dollars where it expects a more rapid payoff. Theoretically, say Weyerhaeuser researchers, tree crop yields could be increased twenty-two times above those of a natural stand using a panoply of high-yield techniques demonstrated in the laboratory. The full range of today's off-the-shelf technology, including irrigation, could bring a fivefold increase. Given current economic realities, the company will settle for a respectable twofold boost in production in the near future.

Contemporary silviculture based on technology pioneered in food production. Impressive gains have been realized in this century through mechanization, breeding, and the use of fertilizers and pesticides. New varieties of wheat produced a Green Revolution that made Mexico and India self-sufficient in basic food production. It isn't clear, though, how sustainable the gains will be either in agriculture or silviculture. "Miracle crops" will continue to grow miraculously only if farmers keep using generous amounts of petrochemicals. The increasing reliance of farmers around the world on a few "elite germplasms" leaves crops vulnerable to massive attack by a single pest. Consider the mysterious blight that attacked the U.S. corn crop in the late 1960s.

In 1968, ears of corn on seed farms in Illinois and Iowa began showing a rot. The next year, the rot had spread not only to more seed fields and hybrid test plots but onto farms as well. At first the problem was misdiagnosed as a combination of the familiar yellow leaf blight and charcoal rot. It turned out to be a new strain of an unrelated fungus. As the wind-spread blight made its way south to Florida, west to Kansas and north to Canada in 1970, it came to be known as southern corn leaf blight. One billion bushels of corn were lost, as much as half the crop in some southern states. Once the problem was correctly identified, the seed industry was able to crank up production of resistant seed. By 1972, the blight was licked and American agriculture was able to wake up from its brief torment. While the successful response to the emergency may have lulled some observers into complacency, others concluded that not all was well in the heartland.

As recounted by Jack Doyle in *Altered Harvest*, the corn blight had found a "genetic window" that allowed its rapid spread. This window was a vulnerable gene in the T-cytoplasm used throughout the seed business to simplify hybridization. A mutation in a fungus was "perfectly keyed" to take advantage of this one gene found from one edge of the North American corn belt to the other. A later study by the National Academy of Sciences called American crops "impressively uniform genetically and impressively vulnerable." Iowa State University pathologist J. Artie Browning likened the genetic homogeneity of the nation's corn to "a tinder-dry prairie waiting for a spark to ignite it." Here's how Doyle summed up the broader problem:

Today, we may be moving toward a high-tech, house-of-cards agriculture worldwide, with genetic engineering at its base; a system in which one monkey wrench or one unforeseen mutation can create enormous problems. Just as the technology of hybrid corn production "went wrong" in 1970, aiding the advance of the corn blight, the agricultural biotechnologies of genes, microbes, and molecules might "go wrong" on a much grander scale in the future.

Tree farming has produced a monoculture not unlike that of the corn belt. The wonderful diversity of the old-growth forest is replaced by a simpler, more uniform system. Western Washington and western Oregon today are largely covered by even-age plantations of Douglas fir. Within a stand, every tree planted is the same species and age. In a region where trees live anywhere from 500 to 1,000 years, the managed forest is harvested and replanted every 50 years. What's lost in the process is more than the structural diversity that makes the old-growth habitat so rich. Also lost is the less-apparent genetic diversity of the trees themselves. The forest industry's technological leaders argue that genetic improvement programs actually preserve the diversity that remains. They point out, for instance, that a natural stand dominated by a few prolific seed producers may have less genetic diversity than the managed stand that replaces it. What isn't mentioned is the fact that over a larger landscape of perhaps half a million acres, the use of only 300 seed trees *does* reduce the forest's already-depleted genetic diversity. While downplaying the problem, Weyerhaeuser hasn't ignored it. The company more than doubled the number of seed trees at its Rochester orchard in order to broaden the genetic reservoir. There isn't any magic answer to the questions about how far clonal forestry can go. In their field trials with cloned trees, Weyerhaeuser scientists are trying to determine how many different genotypes should be used to stock a particular site or a broader landscape. As Cheryl Talbert puts it, "How few is too few?"

When the complex old-growth ecosystem is replaced with a biologically simpler tree farm, the process is fraught with other risks that may not be immediately apparent. That, in a nutshell, is the message that Chris Maser delivers when he talks or writes about forests.

That message isn't greatly appreciated by his employer, the Bureau of Land Management, which has a straightforward view of its mission in western Oregon. That mission is to sell as much old-growth timber as legally possible. Maser, a mammalian biologist and something of a New Age guru, is extolling the value of the very ecosystem that his employer is destroying. Perhaps it's not surprising that Maser has decided to leave BLM in favor of consulting work and his writing projects.

Addressing a crowd of environmentalists, Maser suddenly veers off the subject of forest ecology. He asks them not to view the timber companies and the federal land managers as their enemies. All of society must reach a consensus on our objectives for managing forests, he says. "And we achieve this by being gentle with each other, trusting each other, and dealing with each other with respect. One four-letter word: love. That's something that's missing. We are all desperately afraid of each other in and out of the agencies and we're afraid to work with each other. . . . Until we're honest with each other, we're not going to get anywhere."

The next morning, I visit Maser in his office on the Oregon State University campus in Corvallis. His shelves are lined with books about forest ecology and mammalian biology. On the walls hang quotations from Aldo Leopold, Chief Seattle, Gandhi, and Einstein. The books are representative of the kinds of scientific contributions Maser has made, the quotations of his insistence on broadening our world view. With almost hawklike intensity, he speaks of his fears that intensive forest management will prove a dead-end street. Maser returned a few months earlier from a trip to Germany, where he visited forests suffering from *Waldsterben*, or forest death. The phenomenon is widely regarded as a product of air pollution and acid rain. Maser wonders if we're overlooking the obvious.

"It is worthy of note that decades of scientific research have concentrated on every possible cause of forest decline *except* that it might be the direct result of intensive plantation management based on ignorance of forest processes," he writes in his book, *The Redesigned Forest*. "The forests of central Europe are now dying; in fact, West Germany recently (1986) issued a postage stamp 'save our forests in the eleventh hour.' *And yet the effects of a century or more of intensive management based on short-term economic expediency are seldom discussed* " [Maser's italics].

The European forest has been altered, drastically, in many ways. The original mixed hardwood-softwood forest that covered much of the landscape has been cut down and replaced by even-aged stands of conifers. Soil chemistry is modified by the conversion to Norway spruce--which increases acidity--and by the lack of decomposing wood. Firewood gatherers strip the German forests of fallen branches as well as logs. Wildlife has become rare. Mycorrhizal fungi, essential to healthy tree growth, are depleted. Thirty to forty species of fungus typically are attached to the roots of a Douglas fir. On the spruce of today's German forest fewer than a dozen mycorrhizal fungi may be found, on some trees no fungi at all on half the roots. Research by Maser and others in the Douglas fir region has elucidated the mutual dependence of trees, mycorrhizal fungi, large logs, and the small mammals that feed on fungi. Isn't it reasonable to hypothesize, Maser asks, that the European forest's decline has something to do with disruptions in those kinds of ecological relationships and the resulting changes in soil chemistry and the below-ground environment? If the United States continues to follow the

European model of forestry, isn't it a matter of time before our forests become equally unproductive?

"What does it take to have sustainable forests?" Maser asks, fixing me with his vivid gaze. "Without sustainable forests we cannot have a sustainable yield of wood fiber. We're focused on the wrong end of the forest. We literally do not see the forest for the trees. We focus on the trees, not the process that produces the forest of which the trees are a part. The forest is one system that simultaneously produces a multitude of products. Trees are only one product. It takes the interaction of all the other products to produce the trees. We can choose to ignore the other products, and I guarantee you that we will alter the forest to the point it will not be sustainable at some point."

Maser isn't the first to suggest a link between intensive management and plummeting forest productivity. Richard Plochmann, a University of Munich professor and former district chief of the Bavarian Forest Service, commented in 1968 on German forests' "well-known and frequently observed" decline over the course of a century. He put the loss of productivity at 20 to 30 percent after two to three rotations of a Norway spruce monoculture. "Our forestry will be carried on even under bad economic conditions," Plochmann wrote:

> We could better the return if we would be willing to give up the high intensity now maintained or if we gave up the principle of sustained yield. We cannot do both and do not want to do either. The first seems imperative for the multiple uses of our forest and the second for the benefit of following generations. . .

Corporate executives and public officials aren't eager to hear messages like those of Plochmann and Maser. The politics and economics of Western culture are driven by the idea of never-ending growth, not by talk about scarcity and limits. In Germany and in parts of the eastern United States, where forest productivity also has begun to decline, it's easier to blame outside factors such as ozone and acid rain than to consider the cumulative effects of intensive forest management. One German scientist attributes *Waldsterben* to "a mysterious virus imported from Czechoslovakia." In the Pacific Northwest, where the conversion of virgin forests into plantations hasn't yet ended, it's too early to detect any declines in the productivity of second-growth forests. Maser's warnings can be dismissed as unproven hypotheses. Who wants to look for bad news?

And so we push full-steam ahead with our brave new forest--"marching ahead," Maser wryly comments, "as though we know what we are doing." The forest-products industry's intensive management practices are yielding some impressive results in today's timber harvests. Almost certainly, the gains in the next rotation will be impressive. Beyond that, when Weyerhaeuser and Georgia-Pacific are owned by the grandchildren of today's shareholders, the consequences are less clear. Some experts are dubious about our ability to maintain production at present levels, much less sustain the twofold or fivefold gains that researchers in the forest-products industry are anticipating. Consider the arguments of Roy R. Silen, a forest-geneticist in the research branch of the Forest Service. In an important theoretical analysis of intensive forest management, Silen points out some of the practical limits to ever-greater timber yields. He does this by describing the techniques used in the Green Revolution to increase agricultural productivity and then examining the applicability of these methods to forestry in the Douglas fir region. The strategy that has boosted food production in the Third World relies on a combination of site improvement and selective crop breeding. Using one technique without the other is a futile exercise, Silen explains.

As Jack Doyle notes, intensive agriculture is fraught with risks. When farmers adopt the new ways, they accept a certain amount of instability. The new hybrid crops, not genetically adapted to local growing conditions, lack the resistance of native crops to pests, diseases, and drought. Farmers come to rely on chemical fertilizers and pesticides that they may not be able to afford. Breeders must continually produce new hybrids to close the "genetic windows" that periodically allow diseases like southern corn leaf blight to threaten a nation's food production. Rarely is a variety of wheat or corn used for more than a decade before it must be dropped in favor of a newer, "improved" variety. The genetically diverse native crops used in traditional agriculture offered the reassurance that there would be at least *some* yield even in the bad years. In the face of a mushrooming world population, however, there has been little choice but to adopt high-tech agricultural practices. The Green Revolution has been one of the resounding successes of twentieth century technology.

How wise is it to apply high-tech agricultural methods in forestry? Clonal forestry is being developed in the Pacific Northwest not to feed a hungry world but to stave off the threat of competition from lower-cost producers in places like North Carolina, Chile, and New Zealand. In some ways, risks are greater in high-tech forestry than in high-tech agriculture, Roy Silen asserts. The forest-products industry routinely begins mass production of seedlings from a few parent trees decades before trees in progeny tests approach rotation age. What happens if serious problems are discovered with a parent tree after its offspring have been planted on 100,000 acres of forestland? Silen worries that genetically improved trees will adapt poorly when they are planted in stands miles from their place of origin, and that fast-growing trees will be less able to resist stress from drought and cold.

Cheryl Talbert believes the results from Weyerhaeuser's progeny tests should put at least some fears to rest: "What we've basically found is that the best families tend to be very robust to the environment. There may be a few extreme environments where one or two of the best twenty-five may drop down into the middle and one or two will come into the middle. But we haven't seen any flip-flops of important magnitude." In other words, the parent trees that grow so well in their early years have proved hardy and adaptable to varied environments as they mature. This finding contradicts so much forestry "dogma," Talbert says, that researchers at Canadian timber giant MacMillan Bloedel were reluctant to report similar results. After Weyerhaeuser researchers spoke publicly about the adaptability of genetically selected Douglas fir, their MacMillan Bloedel counterparts told them, "We were afraid to say anything. We didn't know if we were wrong or what."

Weyerhaeuser's findings, while important, apply to trees younger than twenty years. How will the fastest-growing trees stack up in another twenty-five years, when they reach rotation age? Silen argues for a conservative approach:

> Most past financial evaluations have probably left out, or given far too low a value, to crop reliability. It has been popular to commit forest ownerships to intensified forestry on a vague faith that agricultural successes with a new variety can surely be duplicated. Crop reliability would seem to be the most important requirement of a long forest rotation. It is the first thing given up in applying the strategy to agricultural crops. "Some yield even in the worst years" should be the last thing given up in a strategy for Douglas-fir.

Silen makes two other important points. One is that genetic selection can offer only limited gains unless growing sites are improved simultaneously. Even with site improvement, there is a limit to the gains that can be achieved. Timber yields ultimately run up against the "law of constant final yield." This law dictates that only a certain amount of biomass can be grown on an acre. If this biomass is put into fast-growing "super trees," the acre will support fewer of these than of a more spindly tree. Either way, the total amount of wood fiber is constant. Genetic selection may produce a tree 20 percent larger than the average tree. Less wood will be wasted when these bigger trees are milled, so revenues will rise. But greater total yields per acre depend primarily on site improvement. In the Pacific Northwest, drought and cold limit tree growth even more than nitrogen deficits. There's no way to control temperature, however, and irrigation is far too costly in today's market for wood products. Genetic improvement offers distinct promise in producing a tree of higher quality and a tree that won't fall over from excessive foliage when it's fertilized and irrigated.

Silen's final point--and the most important one--is that we are the stewards of our forests' genetic diversity. The greatest danger, the one "irretrievable commitment" being made to high-tech forestry is the continuing narrowing of the genetic reservoir that accompanies old-growth logging. Native tree populations probably will be needed "for their priceless adaptation":

No landowner could afford the genetic program to reconstitute them, if exact reconstitution was even possible. It would be tragic if long-term costs of stabilizing a less-adapted genetic population were so high as to make Douglas-fir forestry unprofitable. . . . [O]ur native tree populations, which we seem to thoughtlessly waste, may be our prime resource when the world of the 21st century must once again return toward truly sustainable yields.

This is a concern echoed by forest ecologist Jerry Franklin, who questions the wisdom of restricting the genetic base just as the world is bracing for severe climatic changes resulting from the greenhouse effect.

Whatever the long-term effects of genetic improvement and intensive management, some immediate problems already are plaguing the industry. Like the genetically similar grain crops that now girdle much of this planet, the Northwest's monoculture of young Douglas fir offers a tempting feast for insects, mammals, and fungi. The pest problem is compounded when a species is moved outside its natural home range. Douglas fir transplanted to Europe is subject to severe damage by rhabdocline disease and Swiss needle cast. Both of these are present throughout the Douglas fir region of Oregon and Washington, but rarely cause serious problems.

Potentially destructive insects and pathogens are endemic in old-growth forests. But because of the forests' structural and ecological diversity, these organisms and their predators tend to remain in balance. Woodpeckers, which excavate their homes in the large snags abundant in old growth, play a critical role in controlling the population of bark beetles. "Secondary cavity nesters," birds that make their homes in existing hollows of snags, also prey on insects. Chickadees and nuthatches feed on the spruce budworm and the Douglas fir tussock moth. The large snags in which woodpeckers and other cavity nesters live are destroyed along with live trees when an old-growth forest is clearcut.

The fungus that causes black-stain root disease is believed to be native to the Douglas fir region, but scientists only recently became aware of its presence. Why now? Because the fungus is spreading rapidly, with the help of road building and tractor logging. So far, damage has been limited mostly to areas immediately around roads and skid trails, although a few tree plantations have been destroyed. Five researchers from Oregon State University and the Forest Service have warned that black-stain "has the potential, from what we have learned, to kill Douglas-fir on a scale beyond that of any other native pathogen." The fungus is spread by two root weevils and a beetle that flourish in the stumps of clearcuts and precommercial tree thinnings. "Insect vectors add a dangerous versatility to any forest pathogen," the researchers wrote. "When insect activity is triggered by forest management activity as is the case in the Douglas-fir black-stain association the results are potentially catastrophic." Among the scientists' recommendations: minimize soil disturbance and plant either a mixture of tree species or trees other than Douglas fir in affected areas.

Too often, the treatment chosen to stop the spread of a forest pest is to cut down the affected trees. Old-growth stands are cut along with second growth and then replanted with even-aged single-species stands more vulnerable to outbreaks than the original forest. Because pests are naturally present in old growth, these natural forests are sometimes portrayed as a threat to the "thrifty" or "lusty" young second-growth stands on which intensive management is practiced. As in the Vietnam pacification program, the strategy sometimes is to destroy the forest in order to "save" it. *Phytophthora lateralis*, an exotic root disease, is ravaging the Port Orford cedars of Oregon. The spores of this fungus are spread downstream from road cuts, infecting and killing all the cedars in their path. What's the Forest Service's response? Not to modify its road-building plans but to log the cedar threatened by new roads. The tragedy is that Port Orford cedar is the most endangered of all conifers in the coastal forests, at risk of disappearing entirely from the wild. Forest management is hastening its demise.

Virgin forests are being logged or burned in a number of areas, in the name of pest management. From the Rockies to the Cascades, forests that could never be logged profitably are being clearcut to stop the spread of dwarf mistletoe and spruce budworm. Mistletoe, which comes in several host-specific varieties, is as old as the forest itself. Spruce budworm, according to one Forest Service study, improves the commercial value of some ponderosa pine forests by culling out lower-value species.

The red-cockaded woodpecker, already on the decline because its habitat is being logged, now is threatened by the Forest Service's war on the southern pine beetle. Entomologists attribute the beetle's spread to the establishment of loblolly pine plantations outside that tree's native home range. Loblolly pine, an excellent commercial species, is far more susceptible to infestations of the southern pine beetle than are native pines and hardwoods. To control the beetle, the Forest Service is using "tree crushers" and napalmlike flamethrowers to destroy more hardwoods. Destruction of the hardwoods removes one more barrier to the spread of the southern pine beetle. The loss of mature trees and snags also eliminates the red-cockaded woodpecker, a predator on the pine beetle. It's almost as though the government's pest-control strategy was designed to increase pest populations.

Pest problems are inevitable, to some degree, whenever human hands alter the vegetative landscape. The new landscape can be designed either in a way that invites hordes of pests or in a way that discourages major outbreaks. Replacing the diversity of old growth with the uniformity of an even-aged monoculture is a recipe for disaster. Pests can spread like wildfire through the crowns or roots of a plantation of nearly identical trees. In more complex forests--the most complex of all being old growth--the spread of pests is slowed not only by the forest's diverse structure but also by a greater abundance of spiders and birds that function as "generalist" predators.

Oregon State University epidemiologist T.D. Schowalter and Forest Service researcher J.E. Means discuss how the pest problems are exacerbated in even-aged tree farms:

> Current forest management concerns in the Pacific Northwest include root beetles, woolly aphids, gypsy moth, black-stain root disease, and Port Orford cedar root rot, all promoted by road construction and/or young monocultures. These concerns indicate the severity of future problems if we continue to convert landscapes dominated by old-growth forests with their physical and biological barriers to pest activity to landscapes dominated by extensively roaded young monocultures with no impediment to pest activity. . . . Remnant old-growth forests also may become more vulnerable because pest pressure from surrounding young stands will increase as old growth is cut.

An exhibit in the library of the University of Washington's College of Forest Resources displays "ENEMIES OF DOUGLAS-FIR." Among them are canker, termites, and Douglas fir beetle. Somehow, Douglas fir has survived these and other pests over millions of years of evolution. The exhibit doesn't mention the axe or the chain saw. In little more than a century, humans have substantially narrowed the genetic variation within Douglas fir. Its "enemies" were never so effective.

The simplification of a forest begins the moment the chain saw first rips into the virgin trees, taking down everything in sight. In our well-meaning but sometimes short-sighted war on waste, we remove every board foot of merchantable timber. Up through the late 1980s, timber managers in some national forests required loggers to gather all the unusable logs in a clearcut into a single pile. Except for the stack of YUM ("yarded unmerchantable material"), the clearcut would be stripped bare. The environment is simplified further through slash burning. Needle-covered branches, small logs, and woody debris are torched in order to simplify replanting and reduce the danger of wildfire. (One of the early reasons for slash burning no longer exists. When reforestation depended on natural or artificial seeding, it was important to clear off the duff layer; Douglas fir will germinate only in mineral soil. With hand planting, duff won't hurt regrowth.) Burning converts nitrogen to a form more readily available to plants, but that benefit is overshadowed by the ecological damage. At typical slash-burn temperatures, all mycorrhizal fungi and nitrogen-fixing bacteria are destroyed. Nitrogen and carbon are released into the atmosphere. A hot fire may volatilize 95 percent of the soil's nitrogen. Burning also has fallen out of public favor because of the severe air pollution it creates. Yet slash burning continues. Coupled with tropical deforestation, it's a leading source of the atmospheric carbon build-

up responsible for the greenhouse effect. Brazilian scientists in 1987 calculated that land-clearing fires in the Amazon accounted for one-tenth of the
carbon released into the atmosphere by human activities.

After "site preparation" comes the new forest. One- to three-year-old
seedlings--almost invariably Douglas fir--are planted in evenly spaced rows.
Racing to form a closed canopy, the young trees look as uniform and tidy as
so many clones. In a few years, they may *be* clones. The trees of a managed
forest never will be draped with thick layers of mosses and lichens to pull
nitrogen from the atmosphere. Soil-building alder and other deciduous plants
are suppressed with chemicals. Within half a century, just as the Douglas fir
are reaching their adolescence, the trees will be harvested and replaced with
another crop. The forest has been simplified, the process of succession
truncated. It's arguable that this plantation isn't a forest at all. Intensive
management has so altered the plant and animal communities that this stand
of trees bears no more resemblance to a natural forest than a zoo's savannah
exhibit bears to a true savannah. Nature has been subdued.

This is not to suggest there is no place for intensive management. Tree
farms will provide almost all of our lumber and paper in the twenty-first century. By producing more board feet per acre, intensively managed forests are
helping meet the world's demands for wood products without sacrificing the
last of the virgin forests. Tree farms needn't eliminate all attributes of the
natural landscape. Flying over Weyerhaeuser's Snoqualmie tree farm, I've
been impressed by the number of mature trees left along the banks of
creeks--far more trees than required by law--for the benefit of fish and wildlife.
And when Weyerhaeuser plants clones *en masse*, the plan is to minimize the
risk of mass failure by separating patches of genetically identical trees.
However, there's no disputing the fact that the artificially managed landscape
is radically different than the natural forest.

How might forestry be practiced in a way that maximizes both biological
diversity and sustainable timber production? Recent research into the old-
growth ecosystem provides some clues. The key concept, in the words of top
researchers, is "heterogeneity." The most resilient, stable, pest-resistant forest
is rich in its variety of vegetation. Douglas fir is not the only species. Hemlock,
cedar, silver fir, Sitka spruce, grand fir, and yew may be mixed in. Douglas fir
may not be planted at all if a site favors other species. Trees are of varied age,
from the smallest seedling on a nurse log to the grandest forest patriarch.
Large dead trees, both standing and fallen, are present. Wildlife are abundant,
including woodpeckers to eat insects and flying squirrels to spread mycorrhizal fungi to the roots of young trees. The forest canopy is uneven, like that
of a natural forest. Perhaps the single most important element of a more
structurally diverse forest is retention of large trees from one rotation to the
next. These trees provide critical habitat while alive and after they die.

Forest ecologists have proposed two basic approaches to incorporate
living and dead trees into the managed forest. These approaches need not be

mutually exclusive. One is to adopt some sort of uneven-age silviculture. The other is to incorporate old growth into the managed landscape through very long rotations. Let's look at uneven-age management first. The type of forestry practiced today on nearly all the commercial timberland in the Douglas fir region is even-age silviculture. One reason for the dominance of even-age forestry is that it's the simplest and most efficient method available. It brings the assembly line to the forest during harvest and replanting. The other reason for even-age forestry in the Pacific Northwest is the intolerance of Douglas fir to shade. Unlike western hemlock and many other trees, Douglas fir cannot grow in the shade of a thick forest canopy. Even-age silviculture allows Douglas fir to grow uninhibited by excessive shade. As it turns out, though, Douglas fir's shade intolerance has been exaggerated. "It is actually fortunate that the routine of clearcutting, burning, and seeding or planting of Douglas-fir has worked at all," Yale University forestry professor David Smith has said. "In most instances the optimum environment for young Douglas-firs is found underneath partial shade."

Classic uneven-age management, or selection cutting, is practiced in some of the drier forests of eastern Oregon and eastern Washington and California's Sierra Nevada. Loggers cut the most mature trees in a stand, leaving younger trees of varying ages to continue growing toward maturity. A hybrid sort of forestry known as "shelterwood cutting" is used on some sites to shade seedlings from the direct sunlight that can produce excessive soil temperatures on charbroiled clearcuts. One-third to one-half of the mature trees in a shelterwood cut are left standing. After the young stand is firmly established, the protective overstory is harvested. The forest is reduced to an even-age stand.

Forest ecologist Jerry Franklin proposes a modified shelterwood regime as an effective way of incorporating more structural diversity into managed forests. Instead of cutting down the mature trees when the young growth is established, the overstory would be allowed to continue growing. For years, Franklin and other forest scientists have stressed to timber managers the important role that large living and dead trees play in the forest ecosystem. As a result of their work, some Forest Service ranger districts--most notably the Willamette National Forest's Blue River Ranger District--have begun leaving some live trees, snags, and large logs as "carryover" for the new forest. Franklin sees large live trees as the key to this carryover. If some live trees aren't left standing, he notes, there will be no snags and logs in future rotations.

The Forest Service and the forest-products industry have spent much of this century explaining to a skeptical public that clearcutting and even-age forestry are environmentally sensitive because they mimic nature. "What many people don't know is that I'm clearcutting to save the forest--the same way Nature does," a logger informs us in an advertisement widely distributed by the American Forest Institute in the early 1970s:

These trees are Douglas fir. Here in western Oregon, their seedlings are only going to grow out in the open, where they can get
plenty of sunlight, so if I don't clearcut, Nature will--with winds
or disastrous fires that burn out thousands of acres. That's been
Nature's way for 10,000 years, but I can do the same thing by
planned harvesting and regeneration--and the trees I clearcut can
be used instead of going up in smoke.

I've logged areas, burned the slash, replanted and watched beautiful new forests come back, so I *know* what we're doing is right.
And I wish more people understood the reasons behind what they
saw before they cried "forest raper."

Wind and fire do destroy forests. The hurricane of 1921 leveled huge
forest stands on the Olympic Peninsula. Fire, even more than wind, created
the forest openings and exposed mineral soil that allowed Douglas fir to grow.
There's enough truth in the industry's advertising to make a case that loggers
are mimicking nature and even improving on it. Many biologists have shared
that mistaken notion with loggers and foresters.

But the notion is wrong for two reasons. First, nature doesn't replace old-
growth forests with perpetually young forests over millions of square miles.
Second, nature doesn't create a single-species monoculture. Only now are
scientists coming to appreciate another distinction between natural forest
succession and industry's version. The vegetative clean slate produced by
clearcutting is far different from the landscape typically created by wildfire.
The difference is in the amount of living and dead matter carried over from
the old forest into the new.

Rarely does fire strip a forest landscape bare. This point was driven home
to me on a leisurely hike through the magnificent old growth along the
Olympic Peninsula's North Fork Skokomish River. Until I stepped out of the
living forest into a devastated landscape, I had forgotten that fire had raged
through the valley in 1985, two years earlier. Except along the river bank,
the entire south slope and much of the north slope was blackened. On that
autumn afternoon, there was little vegetation to be seen. Yet the devastation
wasn't the only notable aspect of the landscape. Here was a valley full of dead
trees, most of them still standing. Fire had raced through the forest canopy
hot enough to kill the trees but not hot enough to consume much of the
wood. When a new forest grows around the remains of the old, the woods
will be rich with snags and logs.

Had this valley been in a national forest or on private land instead of in
a national park, the charred trees probably would have been logged and cut
into lumber. A new forest, with neat rows of Douglas fir, would have been
established sooner. And it would have been a very different kind of forest than
the one that will grow there. The kind of fire that burns hot enough to kill
whole forests of old-growth trees, as the North Fork Skokomish fire did, is

an infrequent and perhaps atypical event in the coastal region. Rarely do natural forests go through the tidy process of forest succession, starting with almost bare land as industry ads and some textbooks suggest. Even the occasional stand-replacement fire leaves a landscape far more diverse than that of a clearcut.

Clearcutting and plantation-style reforestation make good financial sense, at least in the short run. But to suggest that they mimic the natural process of forest succession is grossly misleading. The point is not that the forest-products industry should or can make its managed forests identical to natural forests. That would render commercial forestry impossible. But with a more realistic view of natural processes--processes that have built the land's fertility over the millennia--it becomes possible to incorporate more natural features into the managed landscape at reasonable cost. If landowners and their accountants based investment decisions on profitability over a time scale of 500 years rather than 50 years or less, the vitality of the soil, the genetic reservoir of trees and animal habitat might be viewed far differently. More selective methods of tree cutting might not seem so naive. A longer-term view, so difficult for corporate managers to take, is what truly sustained yield is all about. If private landowners won't take the long view, there's no reason the owners of the public lands--you and I--shouldn't.

Just how far forestry should, or must, move toward maintaining biological diversity in the managed forest has become the subject of passionate debate. Some concerned scientists believe that preservation of the natural forest ecosystem requires more than a combination of old-growth reserves and more sensitive logging practices. Larry D. Harris and Chris Maser are talking about something almost inconceivable from our short-range viewpoint: forest rotations in the hundreds of years. Harris, bringing island biogeography theory to the Douglas fir region, suggests an "archipelago" of long-rotation islands. Each island would consist of a core of continuous old growth surrounded by commercial forests on a rotation age of 320 years. These long-rotation habitat islands would be connected by "travel corridors" of mature forest. Maser, under the rubric of "restoration forestry," offers a model that recalls the agricultural practice of letting land lie fallow in order to rejuvenate the soil. In his scheme, long rotations would be interspersed with shorter rotations. Two 80-year rotations might be followed by a 400-year rotation, then back to two more 80s. Or the scheme could be less repetitive, for instance, two 100-year cycles, then a 300, two 50s, another 300, and three 30s. The possibilities are almost limitless. Timber framers, who use large-dimension old-growth lumber in their post-and-beam construction, have asked that federal timber managers begin managing sustained-yield old-growth stands for their market. It's an idea the shake and shingle industry ought to welcome as well. There's no reason the public lands can't be devoted in part to a specialty product that would also produce ecological benefits.

There's no getting around the fact that a scheme like Maser's or Harris's would drastically reduce federal timber sales. At least in the short run. If the Forest Service and BLM were to adopt such an approach, it's conceivable that private landowners would look at it as a model for restoring their own depleted lands a century from now. As Maser puts it in his hopeful way:

> We, in western Oregon, western Washington, and southwestern British Columbia have the richest conifer forest in the world and, as yet, the healthiest. We have a benign climate, clean air, clean water, and fertile soils; and we have had a severe impact on our forest only within the last century. The ecology of our forest is also better known than that of any other forest in the world. All this gives us the time, if we are committed and we begin now, to learn how to sustain a forest through restoration forestry.

8

Beyond the Timber Economy

Smoke from a cedar fire clouds the few rays of sunlight stabbing through openings in the longhouse roof. Inside this building of hefty cedar timbers and cedar plank siding, a spirited procession of men, women, and children beats on drums and sings loudly in their native tongue. The Tulalip Indians are bringing to their village the ceremonial "first salmon." In this ancient tradition, common to the tribes of the Pacific Northwest, the first chinook salmon caught during the upriver run is brought to the village and later returned to the water where it will come back to life and rejoin its own tribe.

"He will go back to his village and tell the rest of his salmon people how well he was treated," explains Stan Jones, Jr., the elder responsible for reviving the ritual at the Tulalip Reservation near Everett, Washington. "He will bring them back. Their duty is to be caught by our tribal people here. The salmon means the same to the tribe as the air you breathe. They mean the survival of the tribe."

Today, as a thousand years ago, fishing is the basis of the tribal economy. But in these difficult times, tribal leaders know that the return of the salmon depends on more than the good will of the salmon tribe. It also depends on a host of human actions, including logging practices in the Cascades. The fisheries are in trouble. The once-plentiful chinook salmon runs on the Snohomish and Stillaguamish rivers are so depleted that there won't be any fishing for chinook this year. The Tulalips' ceremonial fish was bought from a fish broker. This ancient celebration of nature's bounty serves as a reminder of the ravages man has visited upon one of the greatest marvels of nature.

Those ravages are apparent on Deer Creek, a tributary of the Stillaguamish. Perched on a low bridge above the creek, Gino Lucchetti, points to a rocky island directly in front of him. A fisheries biologist for the Tulalip Tribes, he explains that four years ago there wasn't any island. That spot was an alder-covered bank before the soil and trees were washed away. The rocky channel to the right of the island is new. Perhaps twenty feet of shoreline have disappeared.

Beyond the alders that line the stream is a plantation of young Douglas fir. This is a tree farm, pure and simple. It's hard to imagine that this ever was anything else. Yet the old-growth forest that once surrounded Deer Creek

162

produced a steelhead trout habitat so rich that Zane Grey once rode the caboose of a logging train and then hiked into the woods to visit this fabled stream. He was not disappointed, as he attested in his 1918 account of that trip. "The forest primeval!" the adventure writer enthused. "How cool and fresh and shady and redolent of cedar that deep canyon! How melodious with murmur and gurgle and roar of water! Then the beauty of this Deer Creek and its environment gave me a sense of sheer, wild, exquisite joy." Grey was struck by the large boulders around which formed deep pools, one of which he insisted "could have harbored a swordfish. . . . Deer Creek was the most beautiful trout water I had ever seen. Clear as crystal, cold as ice, it spoke eloquently of the pure springs of the mountain fastness."

As recently as the 1970s, biologists from the Washington Department of Game continued to characterize Deer Creek as one of the best steelhead streams in the state. But conditions have been slowly deteriorating. Loggers moved into the watershed in the 1930s. By the 1950s, they had cut most of the low-elevation timber and began working on the steeper slopes of the upper watershed. Logging and road building may not be totally to blame, but the reshaping of the creeks in the wake of logging is a phenomenon seen over and over in the Cascade and Olympic mountain ranges. Water and melting snow, without forest cover to hold them back, rush into small creeks. Dirt washes in from roads and clearcuts. The more material that builds up in the bed of the creek, the shallower the creek becomes. Where Deer Creek once had swordfish-sized pools, it is now ankle-deep. Its water has to go somewhere, so it laps at the edge of the creek, forming new channels and washing more sediment downstream. This "braiding" process can occur naturally, but is accelerated by erosion from clearcuts and logging roads.

The situation facing Deer Creek took a sudden turn for the worse in the winter of 1984, when a massive slide began at the toe of a steep slope near the mouth of DeForest Creek. Ever since, the waters of Deer Creek and often the Stillaguamish itself have run muddy. When I visited the washout with Lucchetti in 1985, a week after the first salmon ceremony, the slide presented an awesome sight--a semicircular pit with sheer walls, roughly a hundred feet wide and as long as a football field. For hundreds of yards below the slide and the mouth of DeForest Creek, the landscape recalled the aftermath of the 1980 Mount St. Helens eruption. A few months later, the Washington Department of Natural Resources spent $180,000 filling and reshaping the steep walls of the chasm in a desperate effort to stop the erosion. When officials from DNR, manager of the property, held a press conference at the site to show off their work, they discovered that a storm had undone their work during the night. The washout continued to grow out of control. Two years later, officials continue to watch helplessly.

Rain is driving into our faces as Lucchetti and I cautiously approach the precipitous edge. Rocks falling into the chasm make a continuous clicking sound. We stand in the roadway, a road that now leads to a cliff. When I was

Silt from this massive landslide on DeForest Creek east of Everett, Washington, has hastened the demise of a legendary steelhead trout stream.

last here, two years earlier, the chasm was just licking at the edge of the road. Now it has moved well beyond the road. For the past three years, as much as 200 truckloads of dirt have been washed away each day. Will the slide take out everything above the creek?

The slide began in a stand of twenty-year-old second-growth timber. While experts were wrestling with the question of how to control its spread, Georgia-Pacific was logging a large old-growth stand just up the hillside. While there's no evidence that the slide was caused either by past or current logging activities, the disaster demonstrated that the deep glacially deposited soils in this watershed are extremely unstable. Allow them to begin eroding or to become saturated, and the fishery may pay a high cost. The Forest Service, which manages the upper reaches of the watershed, has suspended its timber sales in the watershed and is working with environmentalists, fisheries managers, and other landowners to control siltation. "Silt fences" have been put on cutover federal land, and logs have been placed in streams to slow the rushing water.

Timber-industry executives, with support from many scientists, contend that clearcuts need not cause erosion. This is true. But removal of old-growth forests can have other, equally drastic effects on the hydrology of a watershed. Probably the most damaging kind of occurrence is what scientists call the "rain-on-snow event." When heavy rain falls on a covering of snow in a clearcut or meadow, an extraordinary amount of water can be released in a short time. It was a rain-on-snow event in 1983 that sent tons of logging debris and mud down the hillside into Lake Whatcom near Mount Baker. More than a dozen homes were destroyed by mud and the debris that filled sixty acres of the lake. Property owners sued DNR, Georgia-Pacific, and Scott Paper for damages--settling for $3 million out of court.

The canopy of an old-growth forest, from the tallest Douglas fir to the lowliest devil's club, holds much of the snow that would otherwise fall on the ground. Rain and snow reach the ground more slowly. The duff and humus on the forest floor also catch water and allow it to filter into the soil. The result of all this is that forests moderate the flow of water just as they moderate temperatures. Without this protection, the high-velocity water that flows from rain-on-snow events can scour creek bottoms, washing away salmon eggs in some spots, smothering eggs with silt in others.

The heavy siltation of Deer Creek has meant the loss of some deep pools that once provided important habitat for steelhead. Clearcuts have eliminated the trees that once shaded the creek and kept it benignly cool. And without the occasional large tree falling across the creek, fish have lost hiding places from predators. The damage on Deer Creek's once-abundant steelhead, coho salmon, and Dolly Varden trout has been nothing less than disastrous. A report based on monitoring by two state agencies, one federal agency, and two Indian tribes says steelhead and salmon runs on Deer Creek "are believed to have declined precipitously in the early 1970s. Their current status is unknown but may be at critical levels." Washington Wildlife Department biologist Curt Kraemer in 1985 described steelhead in the now-shallow creek as looking "like they were panting. They were stressed from the lack of cover." Lucchetti that same summer called the creek "a biological desert," reporting that he could find only "one little insect" on a recent visit.

Much of the damage done to anadromous fish habitat has been the result of ignorance. For years, biologists viewed fallen trees as damaging to runs of anadromous fish. It seemed so obvious: logs presented barriers to the migration of salmon and steelhead trout, sometimes causing streams to shift course. To protect the fishery, biologists worked with foresters to see that streams were cleared of these impediments. The Washington Forest Practices Act of 1976 mandated the removal of all large debris from streams following logging. It turned out that this well-meaning regulation was misguided. During the past decade, biologists have learned that, far from hurting fisheries, logs in streams are helpful. (There are some exceptions to this, the ecological value of a log depending on such variables as its size and its exact location in the creek.)

In many streams, the base of the food chain consists of algae that photosynthesize the sun's energy. But in the shady old-growth forest, the food chain begins with needles, cones, bark, and wood. Algae in sunny reaches of streams account for a much smaller proportion of the carbon supply. One of the important functions of logs in old-growth streams is to hold organic litter in streams long enough to be absorbed in the food chain. Leaving streams clean and tidy after clearcutting may increase the total number of fish, but at the expense of certain species. Researchers have recorded growing numbers of steelhead in some streams after clearcutting, but have seen populations of coho salmon and cutthroat trout plummet at the same time.

Logs produce steplike patterns of pools and riffles that reduce erosion and provide a more complex environment. Sediment and gravel rearrange themselves around large logs in a way that creates deep pools. Those pools provide critical habitat for coho and cutthroat. During winter storms in the Cascades and the Coast Range, quiet streams are turned into violent torrents. Immature fish are washed downstream if they aren't afforded the protection of logs or boulders. In smaller streams, half the fish habitat is created by fallen trees. Today, loggers are told to leave large logs in streams and to leave standing trees on the bank to provide shade and a future source of in-stream logs. In waterways as on land, complexity characterizes the old-growth ecosystem.

For salmon, so dependent on pure water and clean gravel spawning beds, there is no better environment than that provided by old-growth forests. Scientists have learned much in recent years about how damage to fisheries can be minimized during logging. Yet the fact remains that logging cannot be done without some damage to streams and fisheries. The damage in coming years may be greatest on the national forests, where roads are being built and clearcuts made on ever steeper, ever more slide-prone slopes. Forest Service soils specialists have managed to steer logging away from some of the most unstable soils in recent years, but they are under pressure to identify *some* locations as suitable for logging.

Forest Service officials insist the agency is doing all it can to protect fisheries. The Mount Baker-Snoqualmie National Forest, in its 1988 draft management plan, proposed to triple the forest's output of anadromous fish --a worthy goal, but one which some biologists view as biologically unattainable. There also were questions about what was politically attainable. In March 1983, Jeff Sirmon, then the Forest Service's regional forester for Oregon and Washington, pegged the region's need for fisheries habitat improvement at $667,000. He was given a budget of $270,000 that year; the next year the fish enhancement budget was completely eliminated. Meanwhile, the region spent nearly half a billion dollars in support of timber sales.

Gino Lucchetti is worried about the future. Eighteen miles downstream from DeForest Creek, he takes me to the mouth of Deer Creek. The view from the levee is remarkable. Ahead of us is the South Fork Stillaguamish River--clear, wide, deep. Beside us is Deer Creek, running the color of coffee with cream. Downstream of the confluence, the Stilly runs brown.

"I think we're starting to see the tip of the iceberg of the problems that are going to haunt us," Lucchetti says. "There are a lot of roads that aren't being taken care of, a lot of clearcuts that shouldn't have been put in because they're on steep slopes, and areas that were logged too quickly. I think in ten or twenty years you'll see these areas failing like crazy."

At one time, loggers didn't have to pay any attention to fisheries. Today, thanks to the federal courts, Indians have a legal club to protect salmon and steelhead habitat. That clout stems first from the 1976 U.S. Supreme Court ruling that affirmed the treaty rights of tribal members to one-half of the anadromous fish caught in Washington state waters. Following up on that decision, U.S. District Court Judge William H. Orrick, Jr., ruled four years later that the tribes had a vested interest in maintaining the habitat that produces the fish. "The most fundamental prerequisite to exercising the right to take fish is the existence of fish to be taken," the judge explained. (Orrick's decision in Phase II of *U.S. v. Washington* became popularly known as "Boldt II," after Judge George Boldt, whose Phase I decision guaranteed Indians half the catch.)

Tribal fishermen and white fishermen had been at war for decades--occasionally exchanging gunfire--over fishing rights. Orrick's Boldt II decision drew a whole new group of corporate landowners and government utilities into the fisheries issue. Shortly after the decision, Seattle City Light dropped plans to dam eleven miles of spawning beds in a Skagit River

tributary and a federal court ruled that farms in eastern Washington couldn't
be irrigated at the expense of salmon habitat on the Yakima River. Boldt II
set in motion a chain of events that has altered the way forest practices are
regulated in Washington. Seemingly intractable conflicts over natural resources have been resolved through a consensus-building process.

Executives of the largest timber companies, manufacturers, and irrigation districts were alarmed over the implications of Boldt II. The tribes had
been dealt a powerful hand when they were given legal standing to challenge
everything from clearcutting to dams and factories to refineries. In response,
the Northwest Water Resources Committee was created. Its membership
read like a Northwest corporate *Who's Who* : Burlington Northern, Chevron,
Georgia-Pacific, ITT Rayonier, Rainier Bank, Seafirst Bank, Scott Paper, Shell
Oil, and Weyerhaeuser, to name a few. The committee hired an energetic
young attorney, Jim Waldo, to advise it on the next step. Waldo outlined four
alternatives: seek to overturn Boldt II in court, ask Congress to abrogate Indian treaty rights, respond to tribal claims on a case-by-case basis, or negotiate with the tribes. Waldo urged negotiation, and his corporate clients
agreed. Within a few months, the tribes and the industrialists were working
on detailed plans to accommodate the needs of both sides on the Nisqually
and Yakima rivers. They also formed a broad-based group that included environmentalists, anglers, and government agencies to find ways to work on
the siltation problems of Deer Creek.

The growing list of successes achieved by the industrialists and the tribes
led to cooperation in other areas. Bill Wilkerson, director of the Washington
Department of Fisheries, was dismayed that eight years after Boldt I became
law, his department was still spending more energy fighting the Indians in
court than in restoring endangered salmon runs. He asked Waldo to approach
tribal leaders about the possibility of state-tribal negotiations over the state's
salmon management program. "Frankly, my attitude was if they could get
some of the business interests in the room and some of the tribes in the room,
it was embarrassing if we couldn't get some of the fisheries managers in the
room as well," Wilkerson recalled later.

The tribes at first were reluctant to seek accommodation with a
department that for decades had resisted virtually every Indian fishing rights
claim. But the two sides did sit down--and in 1984 agreed to jointly manage
the salmon fishery. The number of Indian-state fishing disputes going before
federal court panels dropped from sixty-six in 1983 to none in 1984. Tribal
and white fishing interests worked together to bring about the U.S.-Canada
Salmon Treaty. Like cooperative management of state waters, the treaty was
intended to stop overharvesting and thus rebuild depleted salmon runs. The
tribes had found their legal victory in Boldt I to be Pyrrhic. "Half of nothing
is nothing," Indians grumbled over the ailing fishery. Finally, after years of
fighting over a dwindling resource, Indians, state officials, and at least some
white fishermen were working together to expand the resource.

Gaining a measure of control over fish catches was only half the battle. Unless the spawning and rearing habitat was restored, the new fishing limits would be useless. Salmon were suffering from the pollution created by pulp mills, pesticide use, automobiles, and a host of other human activities. The maze of lowland marshes and meandering river channels in which salmon once reared had been filled and diked. Dams had blocked off thousands of miles of spawning grounds.

Streams and rivers in the national forests stood out as a fertile area for habitat improvement. On Puget Sound, fully one-half of all naturally reproducing salmon spawn in the Mount Baker-Snoqualmie National Forest. With still more spawning taking place in state and private forests, anadromous fish truly could be called a forest product. And as the Deer Creek landslide demonstrated, water quality problems deep in the forest can foul many miles of salmon and steelhead habitat downstream. For good reason, tribal fishery managers intent on rebuilding fish runs turned their attention to the forests.

The "Timber/Fish/Wildlife" process, or TFW, was born late in the spring of 1986 as timber-industry lobbyist Stu Bledsoe was delivering a speech to the Northwest Indian Fisheries Commission. Bill Frank, Jr., chairman of the commission, interrupted Bledsoe to ask if he was willing to negotiate an end to the bitter conflict over the state's forest practices regulations. Bledsoe jumped at the chance. By midsummer, representatives of the timber industry, tribes, environmental organizations, and the state bureaucracy met for the first time outside the courtroom. Together, they asked whether a more flexible state forest practices code could do a better job of addressing their varied concerns. Environmentalists complained that existing regulations did too little to protect wildlife, tribes that the rules left fish in jeopardy. Timber landowners were unhappy with the new streamside logging rules proposed by the Department of Natural Resources. The timber industry had another, equally important reason, for taking part in the talks. By reaching an accommodation with environmentalists and Indians, they hoped to avoid the costly appeals and public-relations fiascos that plagued the industry.

Not everyone relished the idea of bringing together so many warring parties. Public Lands Commissioner Brian Boyle "had to be dragged in kicking and screaming," in the words of one insider. Why should the Department of Natural Resources, which he headed, turn over the job of writing regulations to a fractious political mob? The answer was that the mob represented almost his entire political constituency. Boyle quickly became an ardent believer in Timber/Fish/Wildlife.

Participants in the timber "summit" set themselves a grueling schedule. They wanted to come up with their own proposal--if they could agree on one--before Washington's Forest Practices Board took action on the proposed riparian logging rules that no one outside the bureaucracy was happy with. Over a four-month period, the TFW Policy Group and their technical subcommittees held 100 meetings, discussing the whole range of issues that divided them. Streamside management was one of the most ticklish issues. The mixed stands of hardwoods and softwoods that typically grow beside rivers and streams are critically important to fish and wildlife. Steering the riparian subcommittee through the conflicting demands of wildlife advocates and the timber industry was its chairman, Jeff Cederholm. Cederholm, a fisheries biologist working for the Department of Natural Resources, had built his professional reputation with a landmark study on the effects of siltation on fisheries.

The study by Cederholm and his colleagues examined tributaries of the Olympic Peninsula's Clearwater River, where construction of DNR logging roads had caused landslides. The researchers found that sediment from the slides clogged the gravel spawning beds of salmon and steelhead, causing heavy mortality of eggs and very young fish. The Clearwater study is widely cited as evidence for the devastating effects of sedimentation associated with logging. Yet a decade after the results were published, Cedherholm believes that the damage from sedimentation is frequently exaggerated. In most cases, he points out, streams flush out excess sediment within a few seasons. Today, he is more concerned about overfishing, destruction of sheltered "wall-base channels" critical to smolting coho salmon, and excessive logging in the riparian zone. While they remain standing, trees along the banks of creeks provide the shade needed to moderate water temperatures. When they fall, they provide what Cedherholm calls "the fiber that holds the fish habitat together." Large logs dissipate the energy of freshets that could wash fish downstream prematurely, they retain spawning gravel, and they maintain the proper ratio of pools to riffles. Repeated logging along stream banks degrades the habitat. "Eventually," says Cederholm, "the stream unravels. I've seen that occur." The Oregon Department of Fish and Wildlife estimates half of that state's salmon and steelhead pools have been lost, with a corresponding reduction in fish populations.

As chairman of TFW's riparian subcommittee, Cederholm's top priority was to ensure that landowners be required to leave enough trees standing to create future pools--but not to require more protection than necessary. Wildlife advocates weren't thrilled with the outcome, which allowed as few as twenty trees to be left standing along a 1,000-foot stretch of some streams. Those subcommittee members felt that land animals require much more forest cover. But the riparian needs of wildlife weren't as well documented as the needs of fish because less research had been done. If TFW was to work, environmentalists and wildlife biologists would have to compromise on the amount of habitat they sought.

"I base my recommendations on known information," Cederholm explained. "I don't base my recommendations on what people think would be useful. They have to have documented evidence of it or else I'm not going to speak on their behalf. I have to have data. Data talks. The only time in TFW an argument would hold up through the entire process was when it was based on documented evidence. Hearsay and intuitive feelings did not hold up through the process."

To the amazement of many observers and the participants themselves, the consensus-building process worked. In the fall of 1987, the Forest Practices Board incorporated the TFW agreement into its regulations. The agreement didn't just propose new rules for stream protection, wildlife habitat, and road building. It also gave landowners the option of departing from regulations if they could offer site-specific plans that provided at least the required amount of protection to fish and wildlife. The process of modifying regulations was made more flexible too. In a process of "adaptive management," TFW participants agreed to work together in gathering data on the environmental effects of logging practices and to review annually how well the new rules were working. The timber companies agreed to go beyond the letter of the new law by setting aside two acres of trees as upland wildlife habitat for every 160 acres they clearcut. They also agreed to maintain snags "where operationally feasible" for cavity-nesting animals.

Timber/Fish/Wildlife created a new way of doing business on the private and state lands subject to Washington's Forest Practices Act. Federal lands, where the bulk of the old growth remains, were not affected. Environmentalists participating in TFW did not challenge the right of private landowners to liquidate their 70,000 acres of remaining old growth. The TFW agreement recognized those acres as being "of critical importance to the private companies because they provide a bridge until their second growth stands are mature enough for harvest." It was a difficult act of political realism for environmentalists to write off the last of the private old growth. There seemed little prospect of winning that battle either in TFW or in any other forum. The old-growth battle would be fought over the public lands. On that battlefield, TFW extracted an important concession from the Department of Natural Resources: TFW participants would take a close look at management of DNR's last large old-growth stand, the Olympic Peninsula's Hoh-Clearwater Block. As the price for peace over old-growth logging on private land, DNR and the timber industry were willing to extend the old growth debate beyond the national forests to state land. The dimensions of the battlefield had been defined by mutual consent.

Two years after the historic TFW agreement was reached, some skeptics still insisted that the process was bound to break down. But the TFW agreement had become law and participants in the process continued to work cooperatively. What made the process work--aside from some strong leadership--was each party's commitment from the outset to make accommodations to the others. The timber industry was willing to modify its practices for the sake of fish and wildlife. Environmentalists and the tribes, affirming the importance of a viable timber industry, were equally committed to finding common ground. Even if environmentalists preferred not to see clearcuts on private land, they recognized that matters would be even worse if timberland was converted to other uses. A commercial forest is better than no forest at all.

Back in the late 1920s, when Ira Spring first hiked the trails of the Olympic Peninsula, the virgin forests occasionally seemed like too much of a good thing. There weren't so many logging roads those days, and it was often a long hike to reach the distant mountain lakes and meadows at the end of the trail. In the years that followed, Spring became an avid mountain climber. "The trees only got in my way."

Spring, author of several popular hiking guides, is a bit self-conscious as he utters the "treason" that he understands why so much of the virgin forest has been replaced with tree farms. After all, he lives in a wooden house. A native of the mill town of Shelton, Washington, he knows it would be hypocritical to oppose logging as long as he uses wood products. But during his six decades of climbing and hiking, Spring has become more and more disturbed over the extent of logging in the national forests. He can accept a certain amount of clearcutting, even along trails. What concerns him is the deterioration of backcountry recreational opportunities as more of the land-scape is given over to logging.

Spring has seen one after another favorite trail shortened or lost altogether as roads have been punched into once-remote areas. In many cases, roads have been built smack on top of trails. Where trails are crisscrossed by roads, Spring does what other hikers do: he drives to the end of the road. The national forests' network of logging roads--already eight times the length of the interstate highway system--is booming at the expense of the popular trail system. In the first three decades following World War II, the miles of trails in the national forests dropped by nearly one-third. This occurred while public demand for trails was growing. During the 1970s alone, use of trails more than doubled to twelve million "visitor days." Trail declines weren't intended by the Forest Service as an attack on recreation. Most trails originally were blazed into the backcountry during the Great Depression and the preceding decades to provide access for fire watchers and other Forest Service employees. Recreation, when it was thought of at all, was an afterthought.

In 1932, there were two miles of trails for every mile of road in the national forests. By 1983, road mileage exceeded trail mileage by a three-to-one margin. The administrative importance of trails had declined in the years after World War II, as trucks and aircraft became the principal means of patrolling the forests. Only in the early 1970s did the Forest Service acknowledge that the trail system had become an important recreational asset and that something must be done to stop its deterioration. The agency set a goal of adding 20,000 new miles of trails (which would produce a trail system 25,000 miles shorter than the peak in 1944). Building roads to support timber sales remained a higher priority than improving the trail system. Between 1975 and 1980, the Forest Service spent only $10 million annually on trails, compared with $212 million on roads. Over the following five years, as trail expenditures plummeted, trail work fell 75 percent short of of budgeted targets while road building exceeded goals by 44 percent.

The timber industry's allies in Congress, meanwhile, insisted on giving the Forest Service even more road money than it wanted. When Republicans gained control of the Senate in the Reagan landslide of 1980, two of the timber industry's best friends in Washington, D.C., grabbed the Forest Service's purse strings. Senator Mark Hatfield of Oregon became chairman of the Appropriations Committee and Senator James McClure of Idaho took charge of that committee's subcommittee on energy and the environment. Even after the Senate went Democratic again, Hatfield and McClure continued to get their way. In fiscal year 1987, the timber state senators managed to boost the roads appropriation to $229 million--a whopping $50 million more than the administration requested.

The spectacle of supposedly tightfisted Republicans throwing money at the Forest Service during the Gramm-Rudman era was puzzling at first glance. Part of the explanation may have lain in the continuing wilderness battle in McClure's home state and a number of other western states. Under the Wilderness Act of 1964, only roadless areas can be protected from logging with a wilderness designation. Every mile of road built into a wild area removes hundreds of acres from consideration as wilderness. The Forest Service in 1981 quietly began punching roads into areas being considered for wilderness under RARE II, the second round of the Roadless Area Review and Evaluation. In Washington, Oregon, and California, where wilderness bills were passed in 1984, new logging roads began to appear on the edge of the wilderness. Although Congress had authorized road building and logging outside the new wilderness, environmentalists interpreted the new construction as a cynical move to preempt future expansion of the wilderness areas. Forest Service officials hastened to explain that roads were being built only to meet timber sale targets, not to interfere with the wilderness review. Timber sales and roads generally were kept out of the RARE I and RARE II study areas for twelve years. During that period, the roaded areas were overcut, and foresters felt a need to spread the impact. But the Forest Service's reassurances about its intentions didn't satisfy everyone. The agency's attempt to control the "spin" on the road issue went awry during 1984 as an knowledgeable source, apparently a Forest Service employee, placed anonymous phone calls to the nation's major newspapers alleging a Forest Service-industry "conspiracy" to preempt future wilderness designations.

The Forest Service's logging and road-building frenzy has given Ira Spring a new appreciation of old-growth forests. For one thing, most of the great lowland forest trails are now gone. Then there's Spring's own maturity. The long-time hiker is more interested these days in slowing down to appreciate "the birds and the bees," as he puts it. Few of the wonders of the

Olympics and Cascades now impress him more deeply than the woods. "You run into areas with the magnificent age-old trees," he says. "You're going to be inspired when you enter a grove of those, just like being in a European cathedral. It's awe-inspiring. Those magnificent ones outside of wilderness should never be touched. There are just not that many in the world."

Just as hikers like Ira Spring have come to appreciate the majesty of old growth, so have economists come to realize that the great woods of the Northwest have an economic value beyond their use as a raw material for wood products. People in growing numbers are willing to pay money to hike through the cathedral forests. Loggers still talk about timber being "wasted" when forests are protected from logging. Yet the economic value of a forest isn't always maximized by cutting its trees. When is a forest worth more standing than cut? The importance of the question seems so obvious, and yet no one has offered a convincing answer. Economists don't even agree on how to go about answering the question.

No one knows, for instance, how much old-growth forests contribute to the tourism business of the Pacific Northwest. Tourists and business travelers spend $6 billion annually in Oregon and Washington. Revenues are growing rapidly and tourism is one of the region's top industries. The world-famous old-growth forests are an important, if unquantified, draw. In a survey of Washington tourists, 62 percent said they had visited parks and recreation areas. Forty-five percent reported camping, hiking, or engaging in "outdoor exploration"--far outnumbering those who said they had gone fishing, boating, swimming, skiing, or had visited spas. Expenditures in Washington on hunting, freshwater fishing, and "non-consumptive" uses of wildlife (such as bird-watching) totaled $876 million in 1986. Backcountry recreation has become big business. Camping and climbing equipment giant Recreational Equipment Inc. (REI), grosses $150 million annually and employs 1,400. REI reports that nearly one-third of Seattle residents are members of the locally based cooperative. Americans spend more than $800 million annually on tents, sleeping bags, backpacks, and hiking boots.

The Washington Division of Tourism asked Californians what would induce them to visit the Evergreen State. David Tanner, an economic researcher for the Tourism Division, summed up the conclusion: "One of the things they told us over and over is they've got all the plastic, artificial, synthetic attractions down there. They want a change, some variety. The reason they would come here to the Northwest is the natural attractions. That's one of the best things we've got going for us." Californians who participated in a series of "focus groups" pointed to lackluster promotion by Washington's tourism industry and said the state lacked a sharp focus or identity. They showed only little awareness of the state's rain forests, but expressed great interest in them. The focus groups produced four recommendations by marketing consultants Cole & Webber for creating a unique image for the state. The first recommendation was to promote Washington as having the only rain forest in North America. The rain forest also showed up in the fourth recommendation, promoting the state's diversity.

Even without aggressive promotion, the Northwest's forests draw tourists in large numbers. Olympic National Park has become the fourth most popular park in the United States, largely on the strength of its grand Sitka spruce forests. Each year it attracts more than three million visitors who drop money into the cash registers of merchants from Seattle to Forks. When Congress was debating the merits of creating Olympic National Park in the 1930s, locals were thinking less about the coming tourist boom than about the billions of board feet of "overripe" timber that would be denied to local mills.

What struck President Franklin Roosevelt, when he visited the peninsula in 1937, was the glory of the old-growth forest and the devastation on the cutover lands. "I hope the son-of-a-bitch who logged that is roasting in hell!" the president exclaimed while driving past the massive clearcuts near Grays Harbor. Roosevelt not only signed the bill creating the national park the year following his visit, but later expanded its size by proclamation. Although clearcuts today are generally smaller than when Roosevelt visited, many tourists utter words much the same as those FDR muttered.

Timber-industry executives maintain that hikers really don't care whether they're hiking through old growth or younger forests. Some of the most popular hikes in the Northwest aren't forest hikes at all. Those trails are at high elevations, weaving in and out of the timberline from one alpine meadow to another. Ira Spring remembers the logging of the Olympic National Forest near the Hamma Hamma River and Lena Lake in 1930 and 1931. "When I went back there a year or so ago," he says, "my God, those trees almost looked like a virgin forest again. You have to look around and find the stumps again to find the things were logged."

Bob Spence, vice president of Pacific Lumber and Shipping, sees no need to preserve more old growth for recreation: "Some of the most majestic stands from a layman's point of view would be a stand that's in the range of 100 to 150 years because they're healthy and their canopies are intact. They haven't started the old-growth decay process." Spence recalls a visit to an old-growth forest near Mount Pilchuck in which mosquitoes "were driving my wife crazy." When they emerged into a clearing, the wind drove the bugs away.

On its own lands, the forest-products industry practices a forty- to sixty-year timber rotation so there are none of the mature forests in the age class that Spence finds so aesthetically pleasing. Even on the national forests, where rotation ages range from 70 years to 120, few stands of commercial timber ever will approach maturity. Tree farms aren't altogether devoid of recreational value, but they offer a very different experience than old growth. Tourists don't travel thousands of miles to see a second-growth forest. They can stay home in Massachusetts or Arkansas or France if they want to walk in a young, managed forest.

Nor do locals find in second growth either the wildlife-watching opportunities or spiritual sustenance associated with old growth. The spiritual side--the least definable and least quantifiable of all values--helps explain the passion and rigidity with which some people defend old growth. Jim Eychaner, a hiker, climber, and former director of the Washington Trails Association, isn't alone in the way he thinks of old growth: "Quite frankly, I'm a spiritually oriented person. To me, being on a trail is like going to church, only better, because it's only me and the natural world. It's like a form of meditation and prayer. When I'm around the old trees, there's a sense of mystery and of unity with the planet."

Environmental psychologist James A. Swan believes the great environmentalists--among whom he counts John Muir, Aldo Leopold, Rachel Carson, David Brower, and William O. Douglas--all had something in common with shamanistic religions. That's the experience of transcending the physical world, of connection with the divine, through nature. For whites no less than the Coast Salish natives who follow the traditional *seyowen* way, Swan sees in the destruction of the forests violence to the human psyche. "You lose your soul--literally," he says. "In the deep waters of the human psyche, the unitive symbol of the connection of heaven and earth, of the center of heaven and self, is the tree. By cutting down all those old-growth trees, which are so sacred and valuable to the people of that part of the world, you take away that sense of self and the numinous sense that affirms that. You can't get the same sense from a photograph that you can from going out and hugging a tree."

Even photographs of natural scenes carry a certain degree of healing power. A University of Delaware study showed that students stressed out from a film on workplace accidents relaxed measurably within three minutes of viewing scenes of trees, birds, and murmuring brooks. How much more psychic healing is done by a walk through an old-growth grove or a trip to a mountaintop? Millions of Americans place a growing value on the last wild places--and are willing to pay for the opportunity to visit them.

Greg Mikkelsen of Fort Myers, Florida, took pen in hand to express his dismay at the destruction of the Oregon forests he had crossed a continent to visit. In a letter to the editor of *The Oregonian* in Portland, Mikkelson wrote:

> I was stunned to learn what a small percentage of the original forest is left, even in the national forests that supposedly belong to us all. How you could have allowed so many of our incredible trees to be replaced by eroding hillsides and second-growth trees is beyond my comprehension.
>
> The claim of so many bumper stickers that they are renewable resources is absurd. It will be hundreds of years before those forests can regenerate to anything close to their natural grandeur. You should feel shame for betraying your country's trust by

allowing this crime, no doubt for the sake of some jobs and to make a few petty men rich.

I now know, and my friends will, what it really is like out there. And soon enough it will get around that just outside the frame of those pretty pictures used to attract visitors is the wasteland you've created.

A Tucson couple, Walter and Dorothy Peleck, had much the same reaction, voiced in a letter to *The Seattle Times*:

After driving through Washington the past week, we think it should be called "The Clear-cut State." The older clear-cuts are bad enough, but the new ones on the hillside are terrible to behold. Washington left us with a feeling of depression, not a good feeling for a vacationer to return home with.

Virgin forests can be a profit-generating asset for resorts and other merchants. The lush old growth surrounding the Breitenbush Hot Springs is one of the business's big draws. Tucked into the Cascades east of Salem, Oregon, the property is completely forested except for a meadow dotted with hot springs. A footbridge across the Breitenbush River offers a stunning vista of ancient forests and, in the background, the peaks of the Mount Jefferson Wilderness. A stay at the retreat center would be woefully incomplete without a hike on the Spotted Owl Trail or the Breitenbush Gorge Trail. The Gorge Trail winds through magnificent stands of western red cedar and Douglas fir, and hillsides covered with rhododendrons. Every year, Ram Dass leads a group hike into the cathedrallike woods that surround the gorge where the river tumbles wildly through a narrow channel formed by two walls of rocks. The gorge graces the cover of Ray Atkeson's book of photographs, *Oregon II*. The Spotted Owl Trail received its name when two spotted owls, startled by trail blazers from the Breitenbush Hot Springs, flew out of their way to a higher roost. The trail was rerouted to avoid a nest with two juveniles.

Offering a busy schedule of workshops and group retreats, Breitenbush Hot Springs hosts 8,000 guests a year. Not all is well in the woods around Breitenbush, however. The property is surrounded by the Willamette National Forest, and logging trucks are removing ever more of the valuable timber. The North Fork Breitenbush River has been heavily cut over. The old growth has been almost entirely removed from its south bank. A narrow strip of trees left to shade the river is gone, washed away by the torrents that, according to Breitenbush residents, have become more frequent with heavy logging. On the north side, acres of trees around two clearcuts were blackened when slash burns got out of control.

Breitenbush's proprietors wonder whether their business can coexist with logging. In the fall of 1981, more than 200 acres along the North Fork were clearcut. When a fierce storm hit during the Christmas season, logging debris was swept from the seven clearcuts into the river. During that "terrifying flood," as one Breitenbush shareholder remembers it, stumps, logs, and smaller debris jammed the river just below the confluence of the North and South forks. The logjam redirected the torrent, cutting away a forty-foot-wide swath of the bank. A Roman Catholic retreat center, Villa Maria, was undercut, pieces of the building floating past Breitenbush Hot Springs. Huge logs crashed into the retreat center's footbridge.

Sometimes, when logging slash is being burned, guests must endure a thick haze of smoke. During one workshop on death and dying, centuries-old trees were crashing to the ground. The retreat center's concerns over logging escalated into a crisis in the fall of 1986, when community members were startled by the sound of chain saws along the South Fork. A logging company was cutting a road through the forest to take logging trucks across the river to Devil's Ridge. The ridge, five miles long, had never been touched. Connecting Breitenbush Hot Springs with the Mount Jefferson Wilderness, it was the longest contiguous stretch of unprotected old growth left in Oregon. Environmentalists, recognizing the value of the spectacular stands of Douglas fir and cedar on Devil's Ridge, had pushed for its inclusion in the Oregon Wilderness Act of 1984. The House of Representatives voted to include the pristine ridge. But Senator Mark Hatfield, in concert with timber interests, succeeded in keeping Devil's Ridge out of the wilderness.

The logging took place only after one of the most intensive sieges mounted in the Cascades by environmental activists. In a widely publicized series of actions, protesters repeatedly blocked the logging road and spent days sitting high in the branches of endangered trees. At one point, an anonymous letter to the Forest Service claimed that metal and ceramic spikes had been driven into trees. Metal detectors showed no spikes were present. The Forest Service and the loggers from Bugaboo Timber Company outlasted the protesters--but only after twenty-two were arrested and the dispute over old-growth logging had become front-page news throughout Oregon.

Ten months after protesters took to the treetops, the U.S. Ninth Circuit Court of Appeals ordered the logging halted. The court found that the Forest Service had erred in failing to consider an appeal of the timber resale by Breitenbush Hot Springs and the Oregon Natural Resources Council. The appellants insisted that the six-year-old environmental assessment for the sale be redone to consider effects on water quality and spotted owl habitat. By the time the appeals court issued its decision, the bridge had been built across the South Fork Breitenbush and twenty-eight acres of the primeval forest on Devil's Ridge had been clearcut. The remainder of the timber sale was cut early in 1989 after a renewed court battle and civil disobedience.

For Breitenbush Hot Springs as for many other businesses in western Oregon and western Washington, the quality of the forest environment is more than an academic question. It may be a question of economic survival as well. With trees crashing down on all sides of the retreat center, business manager Michael Donnelly wonders what will become of his dream that Oregon can become "the Switzerland of the Northwest." Judy Ray Painton, after bringing her two children to Breitenbush and devoting eight years of her life to building the community and business, also worries about the future. "People come here to be refreshed and renewed," she says, "not to see clearcuts, not to feel depressed."

When foresters and politicians carve up the public lands, deciding which should be managed for commercial timber production and which for other purposes, they have no reliable tools for measuring the economic ramifications. The benefits of cutting down a stand of trees are fairly well defined. No one has fully quantified the benefits of leaving the trees standing. Will heavy logging force a city to build a multimillion-dollar filtration system to protect its nearby reservoir? Will commercial and sport fisheries be damaged? Will restaurants, motels, and fishing guides lose money?

"That's the real tragedy in my mind," says consulting economist Ernie Niemi, "that in this debate you have two rooms. One is lit pretty brightly; that's the timber side. The other room is dark. We don't know how big it is." Niemi's firm, ECO Northwest, was frustrated after it was hired by the Washington Department of Natural Resources to determine the economic impact of alternative streamside logging regulations. "The foresters and industry and others have told us what they think will happen now in terms of, say, sediment loadings. Now tell us what's going to happen to fish. The biologists can't even tell us the direction. Some of them will say that could increase the number of fish because it will do A, B, and C. And others will say no, it's going to hurt fish because of reasons X, Y, and Z. It's hard to make informed decisions under those circumstances."

Even if the biologists could tell Niemi how many fish a particular action would produce or destroy, the economic implications still wouldn't be clear. He could quantify the effect on the commercial fishing industry because fish are sold in a marketplace. But salmon, steelhead, and wild freshwater fish are also caught by sport fishermen. Although economists can estimate how much a typical angler spends on fishing, those figures don't reveal the full value of fishing. Would the angler be willing to pay an access fee to the Forest Service or other landowner who now lets him walk to the stream for free? How much would he or she pay? How much would a hunter pay for access? A bird watcher? A hiker?

Some economists use the "contingent valuation method" to approximate the full value of a nonmarket commodity. The sportsman is asked how much he or she would be willing to pay to maintain or expand a resource. Sometimes the user is asked how much financial compensation he or she would expect in exchange for a reduction in the resource. Research suggests that many elk hunters are willing to pay twice what they now pay to hunt. Economists these days speak of other nonmarket values even more difficult to measure. These are called "option," "existence," and "bequest" values. Option value is the worth of maintaining one's opportunity to use a resource in the future. Existence value is the benefit of simply knowing that something exists. And bequest value is the ability to pass a resource on to future generations. Economists recognize that these values exist; they can't offer us any dollar figures.

The Forest Service's insistence on losing money by selling low-value timber while doing little to meet recreational demand upsets economists as much as it does environmentalists. In six national forests near Yellowstone National Park, the Forest Service calculates that recreation brings four times as many economic benefits as timber sales. Recreation produces sixteen times as many jobs as timber sales and at far less cost to taxpayers, according to Cascade Holistic Economic Consultants (CHEC). Yet in all but one of the six national forests, federal expenditures on timber sales dwarf recreation spending. The usual explanation offered by critics of the Forest Service for the dominance of the timber budget is the cozy relationship between agency officials and the industry. This interpretation may not tell the whole story.

CHEC director Randal O'Toole prefers to explain such uneconomic behavior in terms of bureaucratic budget-building. He points out that the Forest Service's recreation-derived revenues go into a fund controlled by Congress. A portion of timber sale revenues, in contrast, are retained by the national forest that sold the timber. Why, O'Toole asked a planner for a Yellowstone-area national forest, doesn't the forest concentrate on the recreational expenditures that would promote the agency's avowed goal of community stability?

> "Organizational stability," was his answer. "The timber managers on this forest know that they would lose their jobs if this forest became a recreation forest. Although we might hire some more recreation managers, they would not be the same people who are now managing the timber. In addition, we probably need fewer recreation managers than the number of timber managers we now have."

O'Toole's unorthodox proposal for putting the national forests on a more economical footing is to "marketize" the Forest Service. The agency no longer would receive appropriations from the Treasury. Instead, it would become a public enterprise in which each national forest would stay afloat by focusing on the most profitable activities. Only by selling the national forests' resources in the marketplace, O'Toole argues, can the resources be managed efficiently. Below-cost timber sales would be eliminated by forest supervisors required for the first time to look at the bottom line. O'Toole anticipates a 40-percent reduction in federal timber sales (that's roughly the percentage now sold below cost). Forty thousand workers in the timber industry and in the Forest Service would lose their jobs when the government stopped subsidizing the industry. But taxpayers no longer would be burdened with the $50,000 cost per timber job in the Rockies. Displaced workers could be assisted with the $2.2 billion annual windfall to the Treasury.

In seven of the Forest Service's nine regions, timber sales would be replaced by a more ambitious recreation program under O'Toole's marketization plan. Only in Region 6 (Oregon and Washington) would the commercial emphasis continue to be on timber production. Because of the great quality and quantity of timber in the old-growth forests, Region 6 already accounts for more than 40 percent of the Forest Service's timber sales. That percentage would rise dramatically under marketization. Yet even in the Northwest, O'Toole contends that more backcountry recreation could be accommodated by ending timber sales on economically marginal sites. On much of the Mount Hood National Forest, for instance, the Forest Service could make more money selling recreation permits than timber.

Reaction to O'Toole's proposal has been passionate, and in some cases surprising. Andy Kerr, conservation director for the Oregon Natural Resources Council, at first rejected marketization out of hand, fuming, "Some things you just don't put on the market." But after watching Congress take away citizens' right to appeal some timber sales, Kerr changed his mind: "I've concluded that I'd rather have the environment dependent on the rational marketplace than an irrational Congress." Ernie Niemi says it "frightens me" to think of the national forests being managed on the same profit-making basis as private land. If a national forest supervisor is to make the same kinds of decisions as a Weyerhaeuser or Scott Paper timberland manager, there won't be much old growth left to bequeath to future generations. The spotted owl, marbled murrelet, and other old-growth creatures would be placed in even greater danger than at present.

Marketization is appealing for its elegance and simplicity. Some degree of marketization could serve as a much-needed corrective to the Forest Service's disregard for economic realities. Yet the agency's stewardship over the public lands, tempestuous and misguided though it has sometimes been, has not altogether failed. We can thank the response of the Forest Service and Congress to some nonmarket values for the fact that a portion of the old-growth ecosystem will survive beyond the 1980s. The effects of bottom-line, marketized forestry can be seen in the logged-off Great Lake states, the fast-disappearing tropical rain forests, and the Douglas fir monoculture of the Northwest's industrial tree farms. Forest conservation has taken on a new urgency since the days of Gifford Pinchot. This is no time to surrender the national forests to a profit-driven market. A Forest Service even more sensitive to the nonmonetary values of the old-growth forests is needed if the agency is to set the standard for ecologically sensitive forestry.

While O'Toole's prescription for a reformed Forest Service leaves much to be desired, his critique of current management is convincing. As he points out so devastatingly, current policies inflate the agency's budget without serving either the environment or the pocketbooks of taxpayers. Timber sales serve an industry of declining importance in the regional economy. By the late 1990s, dispersed recreation on the Mount Baker-Snoqualmie National Forest may produce more jobs than timber sales currently produce, according to official figures. When ski resorts and other "developed" recreation are considered, recreation jobs already top timber industry employment by a four-to-one margin. The economy of the Pacific Northwest is changing fast. No matter what the Forest Service does, jobs will be lost in the timber industry due to increased automation and depletion of timber inventories on the industrial forests. Pristine forests will increase in value, both in their commodity-generating capacity and in some nonmarket senses. What will happen to the tourism and fishing industries as the most rugged forests are cut? Will high-tech industries choose to locate in Oregon's "Silicon Forest" after the real forest has been clearcut?

With increasingly mobile employees and industries, economists Ed Whitelaw and Ernie Niemi point out that the Northwest will pay some very real penalties for environmental degradation:

> Every worker in Oregon receives, in effect, two paychecks: one denominated in dollars and the other in the state's clean air, clear streams, scenic vistas, publicly owned beaches, and forested mountains. Take away or damage this environment, and many workers will leave, while those who remain will experience a sudden downward shift in their standard of living and will demand additional monetary wages to compensate them for this loss. In either event, Oregon's economic competitiveness will be reduced.

Some towns have seen the writing on the wall, and successfully rebuilt their economic base. Leavenworth, a one-time timber and mining town on the east side of the Washington Cascades, adopted a Bavarian alpine motif and went aggressively after the tourism trade. Today, business is booming. But tourism is viewed with ambivalence in timber country. One lifelong Darrington logger believes that Winthrop, Washington, was "just ruined" when it took on attributes of a resort town after creation of North Cascades National Park. The rugged Tarheels of Darrington are hanging on to their accustomed way of life as long as they can. Yet even they are beginning to look at ways of diversifying the local economy. After being turned down by the U.S. Navy for a touch-and-go landing field for A-6 fighter pilots (too many steep mountains nearby), the Darrington Town Council set its sights on other possibilities including a camper/trailer factory, a firm that would build housing trusses, and an off-road vehicle park.

Bob Wilder, director of Washington's Interagency Committee on Outdoor Recreation, sees the Leavenworth model as "the wave of the future. . . . With planning like Leavenworth, they can harvest another crop, and that's the tourists, by capitalizing on the natural resources which are in close proximity. The natural resources are the attractants which contribute so much in dollars and cents from tourists."

The traditional one-industry economy based on sawing old-growth timber is coming to an end. If towns like Darrington are to survive, timber will have to become one part of a more diverse economy. In the human economy, as in the ecology of the forest, diversity ultimately will bring greater stability. The towns that will attract the new jobs based on recreation and perhaps some high-tech cottage industries will be those whose landscapes are the most pristine. If the last virgin forests are cut, they will be nothing more than the cedars of Lebanon: a memory of past glory.

9

Battle Ground: Mount Baker

I t hasn't been an easy month for Doug MacWilliams. The supervisor of the Mount Baker-Snoqualmie National Forest has been running from meeting to meeting with staff, elected officials, timber industry executives, environmentalists, the press. He's been explaining the national forest's draft management plan, and none of the major interest groups is thrilled with it. Logging truck drivers are forming protest caravans, environmental leaders are holding press conferences, and Earth First!ers are spiking trees.

"It's like throwing a frog in a pot of cold water and gradually bringing the temperature up," says MacWilliams with a soft lisp. "Then he cooks to death before he realizes the water is hot. I think that's kind of where I am. If you threw the frog in a pot of hot water, he'd probably hop out before he got cooked."

On the basis of the Mount Baker-Snoqualmie's size alone, planning is an immense undertaking. The 1.7-million-acre national forest is not only immense but hosts a wide variety of uses. The forest lies along a 135-mile strip on the western slope of the Cascade Mountains, covering an area larger than Rhode Island and Delaware combined. Within an hour's drive of the 2.3 million people of Puget Sound, the Mount Baker-Snoqualmie is one of the top ten national forests in terms of recreational use. It is also one of the most productive and cost-effective timber producers in the national forest system.

But the fire is burning MacWilliams's metaphoric pot for political reasons. The plan his staff has drafted is the first comprehensive management plan the national forest has ever done. The stakes are high for a wide range of constituents. Forest planners are trying, without public consensus to back them, to accommodate both the sawmills and the constituencies whose interests are best served by forest preservation: backpackers who delight in the cathedrallike majesty of old growth, defenders of endangered wildlife, the fishering and tourism industries, municipal water systems, and Native Americans whose religious rites require untouched cedar groves.

The pot has heated up for national forests around the country as they have attempted to implement the National Forest Management Act of 1976. NFMA requires that the Forest Service prepare a comprehensive management plan for each of the 155 national forests. Prior to the law, the agency had prepared different plans for different resources: plans for timber manage-

184

ment, recreation, mining, grazing, and wildlife and fish management. Now, for the first time, each forest was required to develop a plan that would look at the whole picture. The Forest Service was directed to "attempt to complete" all the plans by September 30, 1985. By 1989, final plans still had not been adopted for all forests.

Planning was difficult in part because the Forest Service was shooting at a moving target. Before much work could be done on the plans, RARE II, the second attempt at a Roadless Area Review and Evaluation had to be completed. Until RARE II was translated into wilderness bills, national forest planners didn't know how large their commercial land base would be. In Oregon, Washington, and California, biologists were warning that the northern spotted owl required hundreds of thousands of acres of virgin timber to survive. The national forests couldn't complete their management plans until the Forest Service chief decided on a plan for the owl's protection. That plan--sure to generate a lawsuit by conservation groups--wasn't adopted until December 1988.

Some interest groups were content to let the process drag on. Because NFMA introduced a new set of constraints on timber harvesting, it was clear from the outset that timber sales would drop in most of the national forests west of the Cascades. The longer the plans could be delayed, the longer timber sales would continue at the old, higher level. The point wasn't lost on the timber industy, which in the middle of the planning process insisted that each plan include a "no-action" alternative. Forest Service staffers grumbled that there was no point cranking out numbers for an alternative that would patently violate the minimum requirements of NFMA. But ranking agency officials ordered them to analyze the no-action alternatives, delaying adoption of the new plans.

When the first draft plans finally hit the streets, the public controversy proved to be even more intense than the Forest Service anticipated. The agency expected no more than 300 administrative appeals of all plans nationwide, roughly two appeals per national forest. Unhappy constituent groups filed 600 appeals on the first seventy-five plans released. One Forest Service administrator called those plans "the easy ones." By 1988, the agency was expecting as many as 1,300 appeals of its forest plans. Controversy wasn't limited to the plans that would govern forest operations for the next ten to fifteen years. Timber sales were drawing an unprecedented number of administrative appeals and lawsuits. The planning process and the timber sale program were facing procedural gridlock.

In a sense, the bitter warfare over the forest plans and the timber sales immediately preceding their adoption represented the storm before the calm. After adoption of the plans, citizens would have a much harder time winning appeals of timber sales consistent with the new plans. Conflicts over forest management would never end, of course. The publicly owned forests had become too valuable to too many Americans for too many different

Top: The forest ecosystem north of Mount Rainier, Washington, has been severely fragmented by clearcuts up to one mile square. Bottom: The clusters of mature trees left on a hillside on the Gifford Pinchot National Forest offer a refreshing contrast to the usual ravages of clearcut logging.

reasons. Yet the forest plans represented a watershed in the history of America's public lands. The planning process would complete what could be the last great division of this nation's forests into wild and managed areas. The Forest Service was deciding which of its 191 million acres would be used for industrial and commercial purposes, and which would be left in an essentially wild condition. Seldom had the nation seen land-use planning on such a grand scale.

The central issue in the planning process, the issue next to which all others paled, was the fate of the old-growth forests. In the Mount Baker-Snoqualmie National Forest, nearly half the forested area outside wilderness was virgin forest in the roadless areas. The Oregon and Washington wilderness acts of 1984 "released" these lands for logging or other uses. The jewels of the roadless areas were those with enough lower-elevation old growth to be of significance to the larger forest ecosystem. Once the fate of these de facto wilderness areas was decided, the decision could not easily be reversed. It would take nature centuries to reclaim the old-growth characteristics of a landscape carved up by roads and a patchwork of clearcuts. And given the way the political winds were blowing, the American public wouldn't look kindly on any future attempt by the Forest Service to invite loggers into an important roadless area withheld from logging under the first round of forest plans. For defenders of the forest ecosystem and for boosters of the timber economy alike, what they would get in this round of planning might be all they would ever get.

Forest Service officials explained that there was nothing final about the first forest plans, and in theory there wasn't. Mount Baker-Snoqualmie Forest Supervisor Doug MacWilliams told a gathering of wildlife biologists not to panic if a forest plan fell short of meeting the needs of wildlife: "It will be revised in another ten years, and those things that we find are weak or wrong with this set of plans can be changed in another set of plans." By then, however, another 44,000 acres of virgin timber will have been cut down and trundled off to the sawmills.

No one could predict with much confidence the effect that amount of logging would have on the specialized flora and fauna of the old growth. Major scientific studies of the old-growth ecosystem had been underway for less time than the lifespan of a single forest plan. Critical questions about the effects of fragmenting the ecosystem remained unanswered. Two years after MacWilliams offered his comforting words, biologists still hadn't found a funding source for the kind of multi-year study needed to determine the degree to which the marbled murrelet is dependent on ancient forests. Draft plans were released before the Forest Service's Old Growth Project held its first major scientific symposium to present new findings on the habitat values of old growth.

Protecting wildlife and plants would be much easier, of course, if that was the Forest Service's top priority. Instead, it was one of the "multiple uses" the agency was expected to accommodate. The impetus for passage of the National Forest Management Act in 1976 was the public's growing concern that the national forests were being managed primarily for commercial timber production. In hopes of creating a more balanced approach to forest management, Congress directed the Forest Service to give equal weight to the major uses of the forests: timber, stock grazing, water quality and quantity, mining, fish and wildlife, recreation, and wilderness. It was not the intent of Congress to let the pendulum swing in the opposite direction from timber production; the idea was to strike a balance among the many uses. When the first round of draft forest plans proposed to reduce timber sales in the Douglas fir region--reversing three decades of timber dominance in forest management--the forest-products industry protested that the Forest Service had caved in to environmentalists. Environmentalists, meanwhile, called for much sharper reductions in timber sales, pointing to scientists' concerns that fragmentation of the old-growth ecosystem was causing it to unravel.

The Forest Service's attempt to satisfy the various constituencies was a thankless, and ultimately futile, task. Just trying to satisfy the varied recreational interests was hard enough. Perhaps the most curious political alignment to emerge in debate over the Mount Baker-Snoqualmie Plan was that between the timber industry and a number of recreation groups. PLUS, the Public Land Users' Society, brought together organizations representing the timber industry, hunters, snowmobilers, off-road vehicle (ORV) enthusiasts, cattle ranchers, rock hounds, and miners. The group advocated "multiple use" in its most extreme form: timber sales should be boosted above the Forest Service's proposal, logging should be allowed in the sensitive mountain hemlock zone, and all trails outside the wilderness system should be opened to horses and off-road vehicles. It was a perspective that pitted PLUS's hunters, anglers, and ORV users against conservation groups that felt the best environment for hiking, fishing, and cross-country skiing is one without the roar of chain saws and snowmobile motors.

The glue that bound the PLUS coalition was its desire for unrestricted access to the public lands. The group represented the backlash against the "unnecessary restrictions and regulations" found in the wilderness system. By standing united, PLUS members reasoned, they could beat back efforts to restrict "multiple use" outside the wilderness areas.

The most puzzling aspect of the coalition was the eagerness of its outdoor enthusiasts to see their playground turned into a tree farm. I asked PLUS co-chairman John Hosford about this. Hosford, a director of the Washington State Sportsmen's Council and executive director of the Citizens Committee for the Right to Keep and Bear Arms, appreciates clearcuts for the browse they offer deer and elk. He doesn't share the enthusiasm of environmentalists for ancient forests: "Old growth is a merchantable log. Old growth will reach a point where it rots and blows over. What good is that? You have a 250-year-old tree lying on its side decaying."

Hosford credits the Forest Service timber program with contributing more to recreation than have all the wilderness bills won by the environmental movement. Wilderness, he explains, excludes the majority of citizens who prefer car camping or a Sunday afternoon drive to a strenuous hike. "A perfect example is my own parents. My father will be 78 next month. My mother is 72, suffers from diabetes and phlebitis. She can't walk thirty feet unaided. My dad has an older Chevy Camaro. He'll put together a little picnic lunch and they can drive up into the woods and drive up these logging roads. They can see wildlife and scenic vistas they once wouldn't have been been able to see because it was totally timbered before. Somebody logged it off and now they can see for miles and miles. They can see Mount Rainier and Mount St. Helens, where they couldn't have seen them before."

Caught between the clamor raised by those determined to save the old-growth ecosystem and those seeking access for logging trucks and weekend picnickers, the Mount Baker-Snoqualmie Forest proposed to do the same as the other national forests in the Douglas fir region: divide the old growth between the two warring sides. Roads would be punched into two-fifths of the roadless areas; the rest would remain wild. Timber sales would be reduced enough to lessen the dominance that the timber program enjoyed over the previous three decades. The total amount of land devoted to commercial timber production would be increased, but by a more modest amount than anyone would have expected from the Forest Service.

If the forest plan was a delicate compromise, it was one that left all parties grumbling--and prepared to take their case to the federal courts or to Congress if necessary. It was a successful plan, according to the theory that says, "Everybody's mad at me so I must be doing something right." If nothing else, the planning process clarified the resource tradeoffs implicit in land allocations. The plan showed, for instance, how timber-dependent communities would be affected if the forest departed from the even-flow policy.

One of the alternatives studied in the planning process used the same land allocations as the Forest Service's preferred alternative but departed from the nondeclining even-flow principle. For the first decade of the plan, the allowable sale quantity would drop by only 9 percent, as opposed to the 16-percent drop in the preferred alternative. The departure from sustained yield would just put off the reckoning. In fifty years, depletion of the old growth would force the annual sale quantity to tumble 35 percent below the present level.

That prospect doesn't sit well with MacWilliams, who has seen the results of rapid old-growth liquidation by private landowners. "My hometown of Rainier, Washington, is an economic disaster area. There isn't a single mill operating. When I was a kid there were three or four. The timber is cut out and they're waiting for the next crop. From an economic standpoint it makes sense for the corporation to do that, if their concern is return to their stockholders. From a community stability standpoint it doesn't make much sense."

But MacWilliams says he isn't eager to dictate the forest's broad management policies. That impetus should come from the public: "My perspective and the one I preach on this forest is that we're professional managers and we can manage this forest any way that our owners, clients, want us to do within reason. Our responsibility is to point out to them what the opportunities are, what the capabilities are, what the costs and tradeoffs and the benefits might be in a range that they can understand. Then once they tell us what their management objectives are that will fit within the capabilities of the land and are legal, we can do that.

"I can manage this forest as a tree farm or a wilderness. I have the capability to do that, and it's no skin off my nose. I don't own it. I own the same share that you do."

The problem is that the public doesn't speak with a unified voice. Instead of a consensus, public input on the national forests' plans is a cacophony of conflicting views. One reason for this may lie with the Forest Service's "black-box" method of planning. Citizens are asked for their views at the outset of the planning process and are invited to offer their suggestions and criticisms of draft plans. On the Mount Baker-Snoqualmie, interest groups were even invited to make management proposals that would then be publicly debated. But the decisions still were made behind closed doors by Forest Service professionals. Everything was public except the decision-making.

Among the critics of this process is University of Michigan professor Julia Wondolleck, who writes:

> This encourages negative, adversarial behavior as each interest seeks to convince the agency decision makers of the legitimacy of their concerns, as opposed to those of their adversaries. It seldom provides an opportunity to focus on the key *issues* of concern rather than those *positions* that might best influence agency decision making. It provides few opportunities for collaboration between competing groups.
>
> Because the process is structured only for comment and then criticism and *not* for constructive development of ideas and solutions to problems, people do not have to grapple with the very real budget, labor, and resource constraints that confront the Forest Service, or with the concerns and interests of other land users.

Wondolleck believes the tools for building consensus are available, if only Forest Service officials would use them. Numerous agency personnel have been trained in conflict management and negotiation since 1982, yet use of those techniques remains the exception rather than the rule. Wondolleck points to examples of successful conflict resolution at the ranger district level in support of her belief that the same thing could take place at the national forest level. Among her examples is the Willamette National Forest ranger district that used consensus-building workshops to develop a five-year timber management plan for an area with heavy user conflicts. Another excellent consensus model, outside the Forest Service, is Washington's Tim--ber/Fish/Wildlife process. In those negotiations, environmentalists and timber companies worked together to produce a new forest practices code.

Richard W. Behan, forest policy professor at Northern Arizona University and a vice president of the American Forestry Association, sums up the planning problem nicely:

> We have witnessed the fate of "preferred alternatives" determined this way: they are preferred by scarcely anyone except the managers. So the least tolerant element in the constituency hauls the managers off to the appeals process, or into court, etc. etc.

As an alternative, Behan advocates what he calls "constituency-based management" of the national forests. Instead of "consulting" with constituents, forest planners would ask them to join in as decision-makers in every phase of the process. The process would be open to anyone "who wants to play the game." Here's his vision:

> All the steps in the management process--problem identification, data collection, analysis, alternative formulation, and choice--are open to participation. Public involvement becomes a continuous process, no longer a series of discrete events.

> Indeed, a negotiated compromise among the various elements of the constituency will almost always result in a "preferred alternative" that proves invulnerable.

The stakes are so high in the conflict over old growth and the roadless areas of the Northwest that it's possible a consensus-building process would fail. Environmental groups or the timber industry might prefer to duke it out at the national level--which is where the process is now headed--rather than solve it at the level of the national forests. Still, it wouldn't hurt to try a new way.

By the time the draft forest plans for the Douglas fir region were out, it was clear that the Forest Service's insistence on making all the decisions made a national struggle inevitable. Since the interest groups hadn't been bought into decisions on matters of central importance to them, the planning process assured that the issue of old growth and the roadless areas would be elevated out of the Forest Service's hands. The dispute was headed toward Congress, with a detour by way of the courts.

Anxiety was growing all the while in timber country.

"City logger!"

A large hulk of a man towers over our table, joshing his old friend. Carol's

Coffee Cup, a blue collar café on the Mount Baker Highway near the small hamlet of Deming, is just where you would expect Bill and Dick to meet again two years after Dick Whitmore left the small sawmill where they met. The walls at Carol's are covered with photos of logging trucks carrying large timber, symbols of this small town's past and present, possibly its future.

The two men briefly catch up on each other's lives before Bill goes back to work. Whitmore, timber buyer for Mount Baker Plywood, resumes his conversation with me. He's explaining why he's worried about the future of independent sawmills like his and Bill's. The timber supply is tightening up, due in large part to the national forests' plans to reduce their sales to the mills. Nineteen of every twenty logs Mount Baker Plywood buys come from the national forests, most from the nearby Mount Baker-Snoqualmie National Forest.

"We used to buy almost all of our wood out of this drainage [the Nooksack River]," he says. "Now I go to Forks, Lake Wenatchee, Enumclaw, to buy my wood. I'm stretched out as far as I can economically bring wood to this mill. If the Mount Baker-Snoqualmie Plan knocks us down another thirty-four million board feet, what's a guy to do?"

Mount Baker-Snoqualmie Forest officials have proposed to reduce the "annual sale quantity" from 204 million board feet to 170 million--a 16-percent reduction. Total timber sales, which include salvage sales of fallen trees, will drop 13 percent, from 220 million board feet to 191.

A visit to Mount Baker Plywood doesn't suggest a company in trouble. The yard is full of old-growth hemlock logs, the lathe is peeling sheets of veneer from eight-foot lengths, workers are pulling veneer off the "green chain," and the layup crew is gluing those sheets together. The plant is running day and night to fill orders for its specialty product, cabinet-grade plywood with hardwood faces. Most telling is the mill's lobby, where job applicants are filling out forms or waiting for interviews. The receptionist is busy scheduling job interviews. For some job-seekers, Mount Baker Plywood is especially attractive because it's a worker-owned cooperative. Production-line workers hire their bosses and receive any profits from the $2 million to $3 million in monthly sales. The hiring boom at this 250-worker plant and other mills in this far northwest corner of Washington is a welcome sight in a decade when layoffs have been the order of the day.

The industry has come back up to the top of the roller coaster. Since the early 1980s, when producers found themselves caught between plummeting demand and impossibly high-priced timber contracts, 100 Washington and Oregon mills have closed their doors. Only the efficient survived. In 1987, Oregon and Washington churned out more lumber, with fewer employees, than ever before. Wages have gone down and production lines have speeded up. Industry profits are impressive once again.

But with timber sales soon to drop on the Mount Baker-Snoqualmie National Forest, Whitmore is worried about the impending descent of the roller coaster. Whitmore decries the "sandwich effect" facing sawmills like his. One face of the sandwich is the loss of private timberlands through road-building, power line construction, and urban sprawl. The other is the reduction in timber sales by the national forests.

Whitmore is an organizer of the Working Forest Alliance, an ad hoc industry group that is warning of the economic repercussions that will follow in the wake of reduced federal sales. In 1984, 27 of Puget Sound's 134 sawmills and plywood mills depended on the Mount Baker-Snoqualmie National Forest for at least two-thirds of their timber. Those mills, together with logging, pulp mills, and furniture plants, accounted for 25,000 jobs in Puget Sound communities. The timber industry represented a modest part of economic activity in the larger cities, but dominated the economy of smaller communities such as Deming, Granite Falls, and Darrington. The Working Forest Alliance calculates that the Mount Baker-Snoqualmie Forest Plan will result in a loss of 1,239 jobs and $30 million in annual wages. The Forest Service projects only a loss of 200 jobs and $10 million.

When compared to this year's timber sales, future sales on the Mount Baker-Snoqualmie will drop even more drastically than is suggested by the forest's public-relations statements and the draft planning document. While forest officials talk about a 13 percent reduction in timber sales, the immediate reality facing Mount Baker Plywood and other mills is something closer to a 35 percent drop in timber sales. In the second half of the 1980s, the Forest Service has been offering larger quantities of timber for sale, an amount far above sustainable levels. Demand for timber is running high once again. Timber contracts voided through the buyback legislation of 1984 have been resold along with some new sales. In 1987, the Mount Baker-Snoqual-mie Forest sold an estimated 260 million board feet of timber and purchasers cut a whopping 295 million board feet (some of that from earlier sales). If the Forest Service adopts its preferred alternative, annual sales will be reduced to 191 million board feet. Environmental groups are pushing for deeper reductions in timber sales.

Virtually all of the buyback resales have been sold, so timber sales will drop even if a more restrictive forest plan isn't adopted. What timber buyers like Dick Whitmore will find at Forest Service auctions will be fewer board feet of timber, with the resulting consequences of higher prices and a lower profit margin. As in the early 1980s, only the strong will survive.

"Right now," Whitmore says, "we're looking at each other in the eye: Is it going to be you or is it going to be me? Who's going to be leaner and meaner? Who's going to find a way of making something cheaper?"

Curiously, while environmental groups and independent sawmills were fighting over the timber sales, there wasn't much talk about what kind of tree *growing* the national forest should be doing. Mount Baker Plywood's line of specialty veneers requires old-growth "white woods" (primarily silver fir and hemlock), products the national forest doesn't plan to grow after the present old growth is liquidated. Timber framers, who need old-growth timber to build their rugged post-and-beam buildings, are so concerned about future supplies that the Timber Framers Guild of North America has called on the Forest Service and the Bureau of Land Management to develop a plan for sustained-yield production of old growth.

Second-growth old growth, if you will, isn't in any timber manager's product line at present. Yet it might make sense for the national forests to develop the product. Neil Sampson, executive vice president of the American Forestry Association, is proposing that the national forests develop a sustained old-growth market niche:

> . . . [t]he tree farms of the Pacific Coast are producing more fiber per acre than was imaginable a few decades ago. So are the plantations of the South. But tree farms seldom produce old-growth timber, because it is not economically profitable.
>
> National Forests can provide that high-quality timber, as a unique contribution to the full range of quality wood products needed in the region, without directly competing with the younger products of private tree farms.

During our afternoon together in the Nooksack River Valley, Whitmore and I saw an impressive variety of uses of the forest. On Church Mountain we watched old-growth cedar, hemlock, and silver fir logs being yarded and loaded onto trucks. The watershed is producing a lot of timber, despite the visual constraints on harvesting. Cars passed by us on the way to the Mount Baker ski resort. We saw cross-country skiers taking to a backcountry trail. We marveled at a ten-foot-diameter Douglas fir in a research natural area--a tree whose age Whitmore guessed as "six hundred fifty-plus, maybe a thousand." Back on the Mount Baker Highway, we saw two eagles, an adult and a juvenile, perched on a branch above the Nooksack. Whitmore has made his point that a forest can be managed for multiple uses. Logging and recreation aren't mutually exclusive.

But the long-simmering dispute that blew into full-scale political warfare during the last quarter of the century isn't over the areas in which loggers and recreationists already coexist. It's over the virgin forests that haven't been given wilderness protection or been put on the auction block. Will the old growth be left wild, or will it be opened up to roads and clearcuts? The fate of the roadless areas is to be determined through the national forest planning process.

Some of the most spectacular old growth was never considered for wilderness protection in the RARE II roadless areas study. Consider the Mount Baker-Snoqualmie Forest's Finney Block. Roughly 144 square miles, or 92,000 acres, of forested land lie in this tract northwest of Darrington. The block is managed almost exclusively for commercial timber purposes. A

Forest Service map shows fewer than five miles of trails, versus 170 miles of logging roads. Only recently have roads been pushed into the little-known North Branch watershed on the North Fork Stillaguamish River. This stunning, nearly intact, four-mile-long valley wasn't studied under RARE II because it contains fewer than 5,000 roadless acres.

It's a sunny June morning when I head toward the North Branch. Although it's been warm for several weeks, a clerk in the Darrington Ranger District office warns me that the mountain road leading to the creek still may be closed by snow. I soon discover that the only peril on the road isn't from snow but from the erosion-caused cliff that's crept to the edge of the dirt road just below a large clearcut. The landscape along the North Fork Stillaguamish is a patchwork of clearcuts, second growth of various ages, and a few modest stands of old growth. As the road rises into the steep mountains that separate the Stilly and Skagit river valleys, it becomes apparent that there is a large quantity of timber yet to be cut--or spared--here. But by the time there's as much virgin forest as cutover land, the terrain and forest type has changed considerably. Here, well up the South Branch drainage of the Stillaguamish, it's a higher-elevation forest with more Pacific silver fir than Douglas fir and far fewer board feet of timber than in the valley bottoms.

The ranger district clerk was right: the North Branch road is still deep in snow. But a Forest Service patrolman in a pickup tells me the Middle Branch road has been opened. ("We had to get in there and replant before it starts sliding.") He directs me to a newer logging road that splits off from the Middle Branch and crosses the ridge to the North Branch.

I take the new road, but abandon my car because it won't make it through the snow. As it turns out, the road is impassable after the next bend anyway. Wind has toppled a number of silver firs and western hemlocks across the road. There hasn't been much human activity here since the road was cut, but there are plenty of signs of wildlife. On the road are the fur-rich droppings of a medium-sized carnivore, perhaps a coyote or a bobcat. A few hundred yards farther, a porcupine ambles out of the old growth and heads toward me oblivious to my presence. Finally taking notice of the stranger, it slowly turns around and disappears behind a log pile left by the road builders. The road offers a view through the trees of a nearly untouched valley. The ragged tree line of the ancient forest stretches from one ridge top to the next. *Nearly* untouched. The edges of two clearcuts twist over the opposite ridge. And the road I'm on spells the future of this ridge.

At the end of the road, I plunge down the steep slope into a forest of silver fir and hefty hemlock. Large mossy logs force me to climb rather than walk through the woods. It's a classic old-growth forest with rich and varied vegetation. Huckleberry bushes and young hemlocks abound, an understory of silver fir and cedar pushing up in spots. The bole of a large, long-dead cedar leans crazily away from the hill. A large cavity at its base smells of rotting wood and an odor that brings to mind the strong organic smell of the stairs in a New York subway. I encounter impenetrable thickets of stinging devil's club and of young silver fir packed as close together as the vines in a blackberry patch. A glade is open and flat enough for a church school picnic. Every way I turn I see the big trees that give the forest its basic structure.

Not far down the hillside I encounter the monarchs of these woods: centuries-old Douglas fir, each four to seven feet in diameter. What event in the history of this forest left such a sharp dividing line between the stand dominated by hemlock and this stand dominated by Douglas fir? One giant tree is obviously charred at the base, the sign of a fire too recent to tell the story of the dominant trees.

I haven't bushwhacked my way far from the road before wondering whether humans have set foot here before. Suddenly that question is answered. And so is the forest's future. I see the yellow markers tacked to the trunks of the giant fir: "BOUNDARY CLEAR-CUT."

To those who spend long hours poring over Forest Service maps and timber sale action plans, new clearcuts usually don't come as a great surprise. Charlie Raines is one of those. He probably was born with a forest map in his hands. Whenever the maps first touched his fingers, they stuck. He's been looking at maps at least since he worked for the Forest Service and then during his years as a Sierra Club activist. The map before this serious young man is a Mount Baker-Snoqualmie National Forest planning map that shows the old growth remaining on this huge forest.

Raines is giving the Forest Service high marks for the clarity of the many maps that accompany its draft management plan. Among the national forest's best maps yet--but still not up to snuff.

"You cannot say for certain that the areas colored in green exhibit the characteristics of old-growth forest." First of all, he explains, the twelve-year-old inventory on which the map was based has been updated only to 1984. Four years have passed since then and another 20,000 acres of natural forests have been logged. Even more significantly, much of what the map identifies as old growth is something quite different. Planners used a definition of old growth unlike the ecological definition developed by the Forest Service's Old-Growth Definition Task Group.

The national forest's figures on old growth are based on a 1976 inventory of "large, mature sawtimber." The dominant trees in this category are at least twenty-one inches in diameter at breast height. This definition is more appropriately used to calculate commercial timber volumes (indeed, this was the real purpose of the dated inventory) than to measure the amount of old growth. In the northern part of the forest, the average age of this large sawtimber is variously given as 230 and 245 years; on the southern part, the age is 250 to 259 years. Impressive ages by our human standards, perhaps, but not by the standards of forest ecology. Forests in the Douglas fir region typically begin exhibiting old-growth characteristics between 175 and 250 years of age. Most true old growth in the Douglas fir region is between 350 and 750 years old. That's how long it takes for forests to develop structural characteristics such as a multilayered canopy, diversity of tree species, large living trees, and an abundance of large trees both standing and fallen.

The Mount Baker-Snoqualmie National Forest wasn't alone in failing to do a scientific inventory of old growth. While preparing management plans, each national forest used its own definition of old growth. Three national forests defined as old growth any stand dominated by trees at least twenty-one inches in diameter. One counted all stands 200 years or older, another national forest 250 years or older. The Umpqua National Forest called any unlogged tract old growth. Not one of the national forests used the ecological definition of the Old-Growth Definition Task Group.

The Wilderness Society hired forest ecologist Peter H. Morrison to apply the ecological definition to six national forests west of the Cascades. Using the Forest Service's own data base, he found there were only 1.1 million acres of true old growth where the forest plans had reported 2.5 million. More than 400,000 acres of the remaining old growth had "little biological value" because it been so badly fragmented by logging and road-building. Most of the remaining old growth was at the less biologically productive higher elevations.

Soon after the Mount Baker-Snoqualmie Forest released its draft plan with the requisite fanfare, Morrison reported that only 297,000 acres of the forest met the ecological definition of old growth--less than half the 667,000 acres reported by forest planners. After eliminating stands that had lost much of their biological value due to adjacent roads or clearcuts, 235,000 acres of "uncompromised" old growth remained. In the entire 1.7-million-acre forest, only 56,000 acres of old growth remained at elevations under 2,500 feet, where biological diversity is greatest.

The ancient forests were disappearing rapidly. Timber sales planned for Mount Baker-Snoqualmie between 1988 and 1993 would destroy 5,000 acres, or close to 10 percent of the low-elevation old growth, Morrison reported. There seemed to be ample justification for Morrison's conclusion: "Continued reliance on inaccurate and inflated old-growth inventories contained in the draft forest plans will lead to actions resulting in an end to the ancient forests."

Morrison's findings should not be taken to mean that forests failing to meet the old-growth definition are of little ecological value. Considerable acreage on the Mount Baker-Snoqualmie Forest came close to qualifying as old growth. More than 180,000 acres met some but not all of the criteria. Those also-rans failed the ecological test because they had too few large trees or snags per acre, lacked a multilayered canopy or lacked the requisite number of shade-tolerant understory trees. On the other hand, most of the stands that met the old growth definition were what Morrison called "early old growth," in contrast to the "classic old growth" dominated by trees over

Much of the congressionally designated wilderness in the Pacific Northwest, like this timberline area in Washington's Boulder River Wilderness, is incapable of supporting old-growth lowland-type forests.

300 years or forty inches in diameter. Some stands lacking logs in the requisite numbers or sizes may have been listed by Morrison as old growth because the national forests' timber inventories lacked information on logs.

As Jerry Franklin has noted, there are "degrees of old-growthedness." A stand whose thirty-two-inch trees qualify it as old growth may not offer any greater habitat value than a nearby stand with thirty-one-inch trees. The criteria put forth by the Old-Growth Definition Task Group, arbitrary though they may be, at least established a measurable standard--the best old-growth definition science had to offer. Most of the national forests, in their vitally important planning process, have ignored these criteria. Even if Morrison's also-rans are lumped in with the true old growth, the Mount Baker-Snoqualmie draft plan still overstated the inventory by 246,000 acres.

The also-rans are important because they are mature unmanaged forests with much, though not all, of the structural diversity of old growth. They meet a great deal of the habitat needs provided by old growth. They are the replacement stands that may become old growth a century from now, when some of today's old growth is laid low by fire, wind, or pests. To a large extent the issue is *natural* versus *managed* forests. Old growth is simply the best of the natural.

When Charlie Raines looks at a map of the Mount Baker-Snoqualmie Plan, he despairs of the continuing logging of valley-bottom old growth. "The proof of the pudding," he says, "is not what you do on White Chuck Mountain, it's what you do down in the White Chuck Valley." I remember his words a few months later as I explore the White Chuck. The last unprotected parts of the valley are being prepared for timber sales. The "Dead Duck" sale will take out a stand of large Douglas fir and cedar. A short distance past that sale site, a Forest Service sign draws my attention. "This area planted 1948 following harvesting," the sign proclaims. "Note how fast these trees are growing to form a valuable new forest." Unable to resist, I walk into the valuable new forest. There are a reasonable number of board feet in these woods, entirely in trees of one foot or less in diameter. The striking thing to my untrained eye, however, is the absence of vegetation under the uniform canopy of this second growth. It's a brown forest. Not even shade-tolerant hemlock can tolerate this much shade. Here is the product of even-age forestry.

There is a place for commercial forestry of some kind on the national forests, just as there is a place for old growth. And the Mount Baker-Snoqualmie Forest Plan, for all its shortcomings, is far more protective of the old-growth ecosystem than draft plans in the other national forests of western Washington and western Oregon. Of the 667,000 acres identified as old growth in the Mount Baker-Snoqualmie, one-third is in wilderness and thus legally protected from logging and roading. After removing other areas from the timber base for such reasons as soil instability, difficulty in reforesting, and wildlife habitat needs, forest planners identified 196,000 acres of virgin timber as "suitable" for logging over the next half century. Half a million acres of virgin forest would remain untouched. As Morrison's study showed, however, the Forest Service figures are misleading. The wilderness areas, centered around spectacular glacial peaks, contain only a modest part of the low-elevation forests. The Alpine Lakes Wilderness is octopus-shaped, leaving the most valuable timber available for logging.

In contrast to the Mount Baker-Snoqualmie, where 27 percent of the "old growth" acres were to be logged over fifty years, the Olympic National Forest Draft Plan called for liquidation of 57 percent of its remaining virgin forests. Only 93,000 acres of the 650,000-acre Olympic Forest would escape the chain saw. Losses of that magnitude would likely prove fatal to a geographically isolated spotted owl population, already reduced to fewer than 200 pairs.

Oregon's Willamette National Forest is the leading timber producer in the national forest system. Roughly 100,000 acres of virgin forest would be cut under its draft plan in just ten years. Two-thirds of the roadless areas would be opened to logging, with more than 40 percent of the remaining "old growth" slated to be cut. Peter Morrison found that only 3 percent of the Willamette's true old growth was below 2,500 feet in elevation, reflecting the heavy pace of logging on the "flagship" of the national forest system.

Although the spotted owl's situation was not as desperate in the Gifford Pinchot National Forest as in the Olympic, the Gifford Pinchot seemed determined to catch up in the race toward extinction. By 1981, this 1.3-million-acre forest had reduced its official inventory of old growth forests to a mere 231,000 acres. Peter Morrison's survey showed only 119,000 acres of old growth, and most of that was severely fragmented. The worst was yet to come. Under the Forest Service's preferred alternative, only 59,000 acres of "old growth"--or 4 percent of the total forest area--would remain uncut. What a contrast to the 1.7-million-acre Mount Baker-Snoqualmie, where 501,000 acres of "old growth"--covering 29 percent of the total land base--would remain standing.

Staff officers in the Gifford Pinchot, concerned that proposed timber sale levels weren't sustainable, launched a crash effort to verify in the field the numbers their computers were spitting out. They found problems with the FORPLAN computer model and with inflated timber inventories. Adjusting for those errors and for a revised spotted owl management policy, Gifford Pinchot planners announced that the allowable sale quantity would be reduced at least 15 percent below the level proposed in the draft plan. That adjustment was intended to restore an even-flow timber policy, not to save old-growth forests.

Why was the Gifford Pinchot being denuded of its ancient forests while the Mount Baker-Snoqualmie was retaining a significant amount? Part of the explanation is that much of the Mount Baker-Snoqualmie was in designated wilderness, assuring the preservation of some large blocks of old growth. But that wasn't the whole story. The Mount Baker-Snoqualmie has twice as much old growth outside wilderness than does the Gifford Pinchot. Vast tracts are up for grabs in both forests. Yet one proposes to keep more than half its roadless areas intact, while the other is virtually liquidating its old growth.

The Gifford Pinchot is what foresters like to call a "working forest," one devoted primarily to commercial timber production. Environmentalists speak more bluntly of the "Gifford Pinchot National Tree Farm," suggesting that timber sales so dominate the forest's management that multiple use doesn't exist in a meaningful way. The Gifford Pinchot, a smaller forest, sells almost twice as much timber each year as the Mount Baker-Snoqualmie, and the Forest Service shows no intention of changing that.

The management differences between the Gifford Pinchot and the Mount Baker-Snoqualmie forests undoubtedly are related to their locations. Because the Mount Baker-Snoqualmie is the closest forest to the major cities of Puget Sound, it makes sense for it to place a greater emphasis on recreation than do some other forests. (This is partly offset by the proximity of Portland to the southern end of the Gifford Pinchot.) The differences also have to do with political realities. As National Audubon Society biologist Chuck Sisco observes, the Seattle-headquartered Mount Baker-Snoqualmie Forest "really took the brunt of the wilderness legislation because you had the Brock Evanses and Charlie Raineses and Jean Durnings right there at your doorstep." After the wilderness battles of the 1980s had ended, those prominent environmentalists continued the fight to save low-elevation old growth. Evans, vice president of the Audubon Society, and Durning, Northwest regional representative of The Wilderness Society, were instrumental in bringing together the nation's leading environmental groups to create the Ancient Forest Alliance in the fall of 1988.

For all the high-powered environmentalists on its doorstep, the Mount Baker-Snoqualmie never ceased to be a working forest. It just simply practiced a form of multiple use that gave greater emphasis to the forest ecosystem and primitive recreation. Yet even there, in the national forest richest in wilderness and most protective of its roadless areas, the spotted owl was in danger. Under the Mount Baker-Snoqualmie draft plan, eighty-four spotted owl habitat areas, or SOHAs, would be established. Planning documents suggested the spotted owl would be assured of ample protection by anyone's standards. So much old growth would be left standing at the end of five decades that the forest would be capable, theoretically at least, of supporting a population of 145 breeding owl pairs.

Only for those who read the fine print was it apparent that the situation wasn't quite so rosy. The spotted owl population estimate, a footnote to the plan noted, was based on "maximum habitat potential." In other words, it assumed that owls would crowd in to use every bit of available habitat and it assumed that each breeding pair required 2,200 acres of old growth. By the time the draft plan was released, researchers had found that spotted owl pairs in Washington typically ranged over 3,800 acres. Even with its optimistic assumptions about owl populations, the Forest Service projected a population drop from more than 190 pairs in 1987 to 145 after fifty years. But the spotted owl doesn't use every acre of "suitable" habitat. By the end of 1988, biologists had found only 49 owl pairs in the Mount Baker-Snoqualmie Forest. Nineteen of those pairs were threatened by logging.

What worried biologists most was that fragmentation of the forest was jeopardizing the ability of young spotted owls to move north or south in search of mates and home territory. The dispersal problem was greatest in the twenty-five-mile stretch between the Alpine Lakes Wilderness and Mount Rainier National Park. Land ownership in the Cedar and Green river watersheds was broken into a checkerboard pattern dating back to the railroad land grants of a century ago. Although federally owned land was scattered into checkerboard squares, the old growth on those squares still was being logged off and the Mount Baker-Snoqualmie Plan proposed to protect only a small part of what was left. Surveys by the Washington Department of Wildlife turned up no spotted owls in the heavily logged Green River watershed and only two in the Cedar.

Environmentalists like Charlie Raines argued that the top priority in the forest plan must be protection of the old-growth ecosystem. A second-order priority was improvement of the trail system. The two goals coincided because the best hiking country includes virgin forests unbroken by roads and clearcuts. "The Forest Service manages more trails in our state than any other agency, yet its trails have been treated like a second-class resource," said Raines. The Mount Baker-Snoqualmie's trail system fell from 1,900 miles to 1,400 over a four-decade period.

The Sierra Club and the Washington Trails Association (WTA) proposed to rejuvenate the trail system with two proposals. The Sierra Club's Trails 2000 plan would reopen 250 miles of abandoned trails and build 350 miles of new trails, boosting the forest's trail network to 2,000 miles by the year 2000. WTA proposed that five unroaded areas be turned into "backcountry areas" devoted primarily to recreation on foot or on horseback. The Trails 2000 improvements would be concentrated in the new backcountry areas, which take in some existing wilderness.

Two of the backcountry areas were basically consistent with the draft forest plan, while three were in conflict. The timber industry and the Forest Service were particularly eager to build a bridge to reach the high-volume timber of the Pratt River valley, one of the proposed backcountry areas. The Forest Service's draft plan, in contrast to the conservationist proposal to boost backcountry recreation, would reduce the forest's capacity for recreation in roadless areas to less than one-sixth of projected future demand.

Beyond their recreation proposals, environmental organizations offered a full land-use plan for the Mount Baker-Snoqualmie. That plan, significantly, would be less damaging to the timber industry than would the Forest Service staff-written alternative emphasizing roadless recreational and environmental values. The staff scenario would result in a devastating 60-percent reduction in timber sales while the conservationist scenario would produce a 35-percent cutback. The environmentalist plan was remarkable for its restraint. Along with the large blocks of green (for no-logging areas), the conservationist map showed an almost equal amount of brown (for intensive timber management). Endorsed by the Sierra Club, The Mountaineers, and the Audubon Society, the proposal signaled that those groups were not seeking to bring timber sales to a halt.

The map for the environmentalist alternative was prepared by Charlie Raines. Conservationists made a concerted effort to avoid a "knee-jerk reaction," he explains. "We made an effort to say, 'Where are some of the more important roadless areas, important old-growth areas? If we were to do tradeoffs, where would some of those come?'" Two issues, aside from timber sale volumes, distinguished the environmentalist approach and the Forest Service's preferred alternative. Conservationists arranged spotted owl protection areas in continuous corridors along major creeks and rivers rather than in the isolated stands mapped out by the Forest Service staff. Almost every available acre in the badly fragmented checkerboard lands in the Cedar and Green river watersheds would be left in a natural state under the environmentalist plan.

The other distinguishing feature had to do with visual constraints. The Forest Service's favored plan would log "at less than full yield" around major highways and recreation areas in order to spare visitors the pain of viewing a landscape dominated by clearcuts. Raines calls the visual corridors a "smoke-and-mirrors" policy that makes it more difficult to find enough acreage both for logging and for wildlife habitat. Conservationists proposed to give up the visual corridors along highways.

The forest-products industry, insisting on keeping timber sales at historic levels, chose not to offer its own alternative plan. The industry sought to maintain timber sales at historic levels. The Public Land Users' Society took special aim at the environmentalist alternative, claiming that its inclusion in the national forest's environmental impact statement violated federal law and "prejudiced" the planning process. With that kind of response from the industry and in the absence of any consensus-building process, the possibility of a political compromise was remote.

The industry's hard line reflected a feeling that it already had bent as far as it could without breaking. Sawmills were still adjusting to reductions in the timber supply resulting from the Oregon and Washington wilderness acts of 1984. Independent mills, faced by overcapacity, had no desire to talk about further reductions in the timber supply. It was not clear, however, that a hard-line position served the industry's best interests. Environmentalists, frustrated by the Forest Service's failure to give adequate protection to the old-growth ecosystem, took the agency to court early in 1989 over its spotted owl plan. Supervisor Doug MacWilliams correctly perceived that an environmentalist victory--which seemed likely--could bring the Northwest timber industry "to its knees." That kind of ruling would be followed by a no-holds-barred effort by the industry to persuade Congress to intervene on its side.

The outcome of a congressional showdown would be unpredictable. But the motivation for environmentalists to compromise was dwindling. Time was on their side. Every story that appeared in *The New York Times* or on network news about the old-growth issue worked in favor of conservationists. By the fall of 1988, environmentalists smelled blood. "Quite frankly," the National Wildlife Federation's executive vice president, Jay D. Hair, told a group of activists, "it's time to kick a little ass." The opportunity for environmentalists and loggers to reach accommodation at the regional or national forest level was fast disappearing.

10

The "Billion-dollar Bird"

The fireworks are flying, as expected. Emotions are running high in the hotel meeting room where 200 people have gathered to talk about the spotted owl. Among those who wait their turn to speak are attorneys, biologists, miners, hunters, loggers, and members of environmental groups. The Washington Wildlife Commission is holding a hearing to decide whether to declare the owl endangered species.

The commission hears many perspectives. More than a few who wanted to make sure nobody will try to keep them, their guns, or their mining equipment out of the woods. Members of the Public Land Users' Society, the antiwilderness coalition of outdoor enthusiasts and industrialists, have turned out in force. An industrial forester, who has been working to set up a consensus-building process on the last state-owned old-growth forests, warns that listing the owl as endangered could "drive us all back into our foxholes." An independent tree farmer worries that his kind may be driven into bankruptcy only to protect a "flourishing" bird that needs no special protection. If the spotted owl were truly endangered, representatives of the timber industry argue, then the U.S. Fish and Wildlife Service already would have listed it.

Each side offers answers to the other's arguments. A retired Fish and Wildlife Service biologist charges that his former employer failed to declare the owl endangered in response to political pressures. "That poor owl needs help," says the biologist, saying the next step is a moratorium on old-growth timber sales. Environmentalists urge the Wildlife Commission to base its decision on the biological facts, not on the possible political or economic repercussions of the decision.

The advocates for both sides occasionally find it hard to keep their tempers in check. When the project leader of a joint state-federal owl survey stands up to correct another witness's characterization of his study, the industry representative asks the researcher if he's calling him a liar. When conservationists point out that the broader issue is preservation of an endangered ecosystem, their foes pounce on those comments to argue that they are merely using the spotted owl as a ploy to achieve other goals. "It became obvious this is not a spotted owl situation, this is an old-growth situation," says a National Rifle Association member.

The spotted owl, because of its close association with old-growth forests, *is* a fitting symbol for the fate of the forests. Yet the Wildlife Commission's charge is to focus on this one question: is the owl endangered in Washington? As a reminder of that central issue, Wildlife Department managers have brought along a stuffed spotted owl. It sits in a display case next to the podium. The brown and white bird's wings are spread partway open as if it were about to take flight. A vigorous bird, a deadly hunter, it appears.

A small plaque on the display case tells what's known about the bird. The juvenile left its nest in the summer of 1986. No one knows where it was hatched. Its life ended the following winter, when it was found in the back-yard of a schoolteacher in the Skagit River town of Sedro Woolley. The bird was hanging upside down, its lifeless talons still desperately holding onto a limb. It had starved to death.

That had become the usual fate for juvenile spotted owls in the 1980s. Nearly all the young tracked by biologists in recent years have died after dispersing from their nests. Many, like this one, starved. Others became prey of the great horned owl.

After spending a full day listening to every argument and counterargu-ment that could be raised, the wildlife commissioners vote unanimously to declare the northern spotted owl an endangered species in Washington. The testimony the commissioners found most compelling was that of Harriet Allen. Allen, a diminutive young biologist has been tracking spotted owls for the past six years and has headed the Wildlife Department's owl study for two. She painted a bleak picture of the bird's situation.

The owl's population was low, she reported, somewhere between 500 and 600 pairs in the state. That population apparently was separated from the larger Oregon and California populations by the Columbia River Gorge. Within Washington, Allen said, logging and development had isolated owls in the Olympic Mountains from those in the Cascade Range seventy miles away. Old-growth logging in the Cedar River and Green River watersheds southeast of Seattle had created a "bottleneck" to north-south movement of owls on the west slope of the Cascades.

The loss of habitat wouldn't have seemed so dire were it not for the spotted owl's poor reproductive performance. The Wildlife Department had found only four juvenile owls in a three-year study of forty-six spotted owl habitat areas (SOHAs) maintained by the Forest Service in western Washing-ton. In the Mount Baker-Snoqualmie National Forest, there hadn't been a single case of successful reproduction. Fewer than half the SOHAs were even occupied by spotted owls and some were occupied by their aggressive competitor, the barred owl.

The few spotted owl fledglings to enter the wide world found it an in-hospitable one. As the old-growth forest was carved into ever-smaller pieces, juveniles couldn't find suitable homes. Almost all the juveniles tracked by

Right: With the amount of suitable habitat continuing to decline, this juvenile spotted owl will be lucky to live long enough to find a home and a mate. Bottom: This Olympic Peninsula hemlock forest, ravaged by a 1921 hurricane, isn't old growth but is a natural forest with the kind of structural diversity in which spotted owls flourish.

scientists in Oregon and Washington met the same fate as the bird that ended up in a display case.

By the late 1980s, very little potential spotted owl habitat remained on private land. Those mature and old-growth forests were being liquidated rapidly. If the spotted owl were to be saved, the critical player would be the U.S. Forest Service, which managed three-quarters of the remaining owl habitat in Washington. But, as Harriet Allen explained to the Wildlife Commission, the Forest Service's measures to protect the owl looked like a recipe for extinction. For all the glowing figures released by the Forest Service about the forests' "capability" to support owls in the future, the reality was sobering. Years of surveys by Forest Service and Wildlife Department biologists had turned up 387 spotted owl sites in Washington. Just over half of those sites would receive any kind of protection under forest plans being considered as the Wildlife Commission met in January 1988.

Nowhere in Washington was the spotted owl's situation more desperate than on the Olympic Peninsula, where Wildlife Department biologists estimated the number of pairs at only 100. In the Olympic National Forest's Soleduck Ranger District, Allen said, spotted owls were known to roost in eleven areas. Five spotted owl habitat areas, or SOHAs, had been created--but they weren't where the owls were.

Allen described the five SOHAs on the Olympic Forest: "Only one of those--*one*--has an owl pair inside the zone. Another one has an owl pair on the edge. The two on the northern edge are not even suitable habitat. They may be in a hundred years or a hundred and fifty years. The third one in the middle had a documented nesting pair. The nest area was harvested, then it was made a SOHA! There is a documented pair outside the area. The protection is not there. Ninety percent of the sites with known owls on them are not protected."

The Olympic National Forest was typical of Washington's national forests, not unusual, in locating spotted owl habitat areas somewhere other than around known nest sites. "In summary," Allen told the Wildlife Commission, "we have severe problems with this population due to low numbers, low density, low reproduction, high juvenile mortality, large home ranges, declining habitat, fragmentation, distribution gaps, isolated subpopulations, and inadequate management.

"This species is in trouble in our state. The primary reason is habitat loss."

That the Wildlife Commission even agreed to hold a hearing on the spotted owl suggested both the biological complexity and the political intricacies of the issue. Only a month earlier, the U.S. Fish and Wildlife Service had ruled that the owl was not endangered. Had this been a straightforward biological issue, that decision by the nation's presumed experts in the area of wildlife extinction would have put the matter to rest. What business did a mere state agency have second-guessing the judgment of Fish and Wildlife? Yet here was the Wildlife Department's top spotted owl expert presenting evidence that seemed to show the bird was in dire straits.

Wildlife commissioners made clear that their decision would be based on the biological facts as they understood them, not on Fish and Wildlife's interpretation of the facts nor on all the arguments about eco-nomics or recreation. The commissioners showed little hesitation in reaching a conclusion.

"It scares me," said Wildlife Commission Chairman James M. Walton, to see aerial photos showing the destruction of the spotted owl's old-growth habitat. For commissioner Terry Karro, the evidence all pointed in one direction: "I think we have no choice in the matter, based on the biological information that's been presented to us. Based on the fact alone of the rather alarming reproductive failures and the high juvenile mortality, I think we have to acknowledge that the spotted owl in the state of Washington is an endangered species."

After the Wildlife Commission's unanimous vote to declare the spotted owl endangered, departing Wildlife Department Director Jack Wayland tried to defuse the politics by saying the decision was no different than past actions by the commission to protect other species. "That's the way we approached it. All the bells and whistles were going off. Every indication was that the species was in trouble."

Bells and whistles had been going off in Forest Service offices for seventeen years. In the fall of 1971, scientists at Oregon State University met with representatives of the Forest Service and the Bureau of Land Management to warn that logging of old-growth forests might place the spotted owl and the goshawk in jeopardy. That meeting set in motion a process that eventually made the spotted owl the most controversial bird in North America, if not the world. Research on the spotted owl became almost an industry. Biologists and technicians were hired by the federal government, three state governments, and the forest-products industry to gather more information about the owl's habitat needs. The Forest Service classified the owl both as a "sensitive species" requiring special management and as an "indicator species" to be used as a barometer for the health of the old-growth ecosystem.

The Forest Service adopted one plan after another to protect the spotted owl, each plan lagging behind the latest research on the bird's habitat needs. Each plan established more extensive habitat areas, yet by early 1989 the agency still hadn't slowed its timber sales to accommodate the owl. By then the bird was in imminent danger both on the Olympic Peninsula and the Oregon Coast Range.

The Forest Service's first tentative efforts in the mid-1970s to protect the spotted owl were prompted by the research of Eric Forsman, then a young grad student at Oregon State University. Forsman's first encounter with a spotted owl came long before he went to grad school. A self-proclaimed "bird nut," he had taken an interest in the owl while still in high school. Of all the owls, the one he was most eager to find was the northern spotted owl, a rarely seen bird that early naturalists believed to be associated with dense, old forests. On his bird-watching trips in the forests of the Oregon Coast Range, Forsman looked--unsuccessfully--for the legendary *Strix occidentalis caurina.*

It was only by chance, while living and working at a Forest Service guard station on Oregon's McKenzie River, that he became curious about "barking" sounds he heard in the night. Forsman echoed the call, and soon discovered that a pair of spotted owls were his neighbors. Thrilled over the discovery, he learned enough about spotted owl behavior and vocalizations from that first pair that he could go out and find more spotted owls. Later, with fellow OSU student Richard Reynolds, Forsman would make research trips into the old-growth forests--Reynolds looking for goshawks, Forsman for spotted owls. Two facts soon became apparent to Forsman: first, that the owl was more abundant than generally thought, and second, that it shunned clearcuts and younger forests. Forsman began his systematic surveys of spotted owl habitat use in 1972. Over the next sixteen years, he and other researchers tracking spotted owls made 95 percent of their sightings in old-growth forests or in mixed forests of old growth and mature stands at least a century old.

If the spotted owl was to survive, strong action by the Forest Service and other federal agencies would be necessary. Two-thirds of the remaining spotted owl habitat was in the national forests, one-sixth on land managed by the Bureau of Land Management. By the 1980s, state and private landowners had liquidated nearly all their old growth. Nonfederal owners in Oregon logged off 77 percent of their remaining spotted owl habitat between 1961 and 1985. Eighty-one percent of owl habitat on private and state land in Washington was lost in the same period.

The spotted owl was a top priority of the Oregon Endangered Species Task Force organized in 1973 by the Oregon Department of Fish and Wildlife. Scientists from U.S. Fish and Wildlife, Oregon Department of Fish and Wildlife, Forest Service, BLM, and Oregon State University served on the panel. The members of the task force's spotted owl subcommittee recommended a minimal level of protection. They asked that 300 acres of old growth be retained around spotted owl nest sites selected for protection.

The Forest Service's regional forester and the Oregon director of BLM, unwilling to jeopardize timber sales, issued a joint memorandum rejecting the proposal until more information was available on the owl's needs. Here's how a Forest Service wildlife staff officer later described the response of the two agencies to the growing concern over the owl: "Management of old-growth timber is a sensitive issue, and there was a desire by some to liquidate the old growth and replace it with younger stands. Agency managers were cautious in making decisions that set aside large acreages of old-growth forest until more evidence supported the need."

That evidence came rapidly. In May 1975, Forsman attached small radio transmitters to the backs of eight spotted owls in the Oregon Cascades. By following the movements of the birds, he was able to document that the birds foraged over a wide territory. The least amount of old growth in an owl's home range was 740 acres. The average was 1,686 acres. For the first time, there were numbers to back up the observation that spotted owl habitat consisted of old growth--and lots of it.

After Forsman reported his findings in 1976, the regional forester responded by ordering spotted owl nest sites protected in Oregon, at least until management plans were developed. In February 1977, the Endangered Species Task Force's long-range plan was adopted by the Forest Service. Four hundred owl pairs were to be protected on Forest Service, BLM, and other public lands in Oregon. Each owl would be provided a core area of 300 acres of old growth, plus 450 acres of forest at least thirty years old.

It was a modest, and wholly inadequate, start. An owl would be given one-fifth of its usual amount of old-growth forest, plus a patch of pole-sized timber offering no apparent habitat value. The inadequacy of the protection plan soon became more apparent. In 1980, Forsman tracked three spotted owl pairs in the Oregon Coast Range. The average amount of old growth used by a pair--2,264 acres--was considerably larger than the area used by an individual. The pairs were followed only for four months during the spring and summer. Had the pairs been tracked for a full year, Forsman pointed out, the reported home ranges probably would have been larger.

How did the federal government respond to the new information? By 1980, Washington had joined in the protection efforts, with the new Oregon Endangered Species Task Force broadened into the Oregon-Washington Interagency Wildlife Committee. The restructured committee came up with a new plan: 1,000 acres of old growth was to be maintained within one and a half miles of a spotted owl nest or center of activity. Three hundred acres should be around the nest site, if known. The new recommendation was incorporated in 1984 into the Regional Guide for the national forests in Oregon and Washington. Far better than the 1977 protection plan, the new policy still provided less than half the amount of old-growth habitat that field data showed the birds typically used.

Forest Service and BLM managers could say they were just carrying out the recommendations of the scientists on the Interagency Wildlife Committee. Committee members, however, were under pressure from the land managers not to disrupt timber sales any more than absolutely necessary. That pressure was most strongly felt by committee members working for those managers. Charged by National Forest Management Act regulations with maintaining viable populations of vertebrate species, Forest Service managers came up with a new concept: the "minimum viable population." Forest Service planners said the concept was used only as an exercise to find a safe number above the minimum. But there were concerns that management plans would allow the population to fall below minimum viable levels.

"We're using minimums across the board," Forsman said in 1985, "whether it's habitat or quality of habitat. The agencies tend to use the minimums and then make rules in the plan using that. There's a real problem with that. If you provide any species with minimum habitat, minimum quality, and minimum population, it's unlikely that they're going to thrive. If you go with the minimums, you're asking for it. The risk of failure is high. You don't have any wiggle room."

Environmental activists saw an opportunity in the Forest Service's inadequate spotted owl plan. Here was a lever to force the agency to take more low-elevation forest out of its commercial timber base. The provisions of the National Forest Management Act and perhaps the Endangered Species Act could be invoked to save a helpless bird. Unlike Tennessee's snail darter, it was a feathered, photogenic creature tailor-made for garnering public sympathy. "In my weaker moments, I've said if the spotted owl didn't exist, we would have to invent it," observed Andy Stahl of the Sierra Club Legal Defense Fund. Stahl emphasized that the issue was the old-growth ecosystem, not any single component of it. The spotted owl, important in its own right, became the symbol for a larger issue: how much further would a unique ecosystem be fragmented?

When the Forest Service released its Regional Guide, the National Wildlife Federation and three Oregon-based environmental groups were ready. They filed an administrative appeal of the spotted owl guidelines. Douglas W. MacCleery, the one-time timber industry lobbyist then serving as deputy assistant secretary of agriculture, found the Forest Service's Regional Guide failed to give adequate consideration to the spotted owl. He ordered a supplemental environmental impact statement (SEIS) that would take into account the most recent research and produce a defensible plan for protection of the owl. A special team in the Forest Service's regional office in Portland took more than a year to produce the two-volume draft SEIS. The document, which offered the most rigorous scientific analysis of the spotted owl's situation yet undertaken by a federal agency, was something of a curiosity.

The new Forest Service plan offered more protection to the owl than had any previous proposal. The preferred alternative would set up a network of 550 spotted owl habitat areas, or SOHAs, of up to 2,200 acres each. Only 1,000 acres of the SOHA would be taken out of the national forests' timber base, the remainder given only interim protection to "maintain options." The plan would remove between 313,000 and 690,000 acres of owl habitat from the commercial timber base of 9.5 million acres on the thirteen Pacific Northwest forests within the spotted owl's range. The timber industry portrayed that 3 to 7 percent reduction in the national forests' commercial timber base as potentially disastrous. If the Forest Service gave permanent protection to the full 2,200 acres per owl, the Northwest Forest Resource Council claimed, $6 billion worth of timber would be put out of the industry's reach. The spotted owl came to be known as "the billion-dollar bird."

Pointing to cases of spotted owls nesting in second-growth forests, the industry questioned whether the spotted owl needed old growth at all. Bruce Engel, president of WTD Industries, was outraged when the Forest Service told his subsidiary to go ahead and log a stand of second-growth timber in the Rogue River National Forest after a spotted owl nest was found in the sale area. One Forest Service official reportedly called the owls "surplus." Why, Engel asked, was it necessary to protect the owl's old-growth habitat if there were surplus owls on second growth? On the east side of the Cascades, where the drier climate produces a different kind of forest than to the west, other cases were reported of spotted owls in young forest stands. The Northwest Forest Resource Council argued that the spotted owl needed no protective measures.

The industry would have a stronger case if the number of spotted owl nests in second growth weren't so low. "Those are the exceptions," said Eric Forsman. "For every one of those, you get ten or fifteen--no, more like twenty or thirty--that are in classic old-growth stands." Nearly all of the spotted owls found in "second growth" have been in stands older and more structurally diverse than the even-age stands of the type being planted by the forest industry these days. Many of those "second-growth" stands included old-growth-sized trees and patches of true old growth.

For all of industry's complaints that the Forest Service's 1986 plan gave the spotted owl more protection than it needed, the agency's own biologists suggested that it actually would mean extirpation of the owl from Oregon and Washington. The effects of habitat loss wouldn't be felt immediately. Population projections in the draft SEIS gave the spotted owl a "low" to "moderate" probability of maintaining a "well-distributed population" in Oregon and the Washington Cascades after 100 years, a low probability in 150 years, and low to very low in 500 years.

The owl population would decline faster on the Olympic Peninsula, where its survival chances would be low within 100 years. The owl would have a moderate chance of long-term survival in Oregon and the Washington Cascades only under the alternative calling for a moratorium on old-growth logging and restoration of additional forest habitat. The Forest Service's choice of the less protective alternative could mean only one of two things. Either agency managers didn't believe their biologists' population projections or they were dismissing them as irrelevant since extinction would occur long after the managers had retired. Retirement, it seemed, meant never having to say you're sorry.

The alternative selected in the draft SEIS also flew in the face of the latest field research on the spotted owl. Choosing once again to err on the side of timber sales, Forest Service managers had decided to give spotted owls far smaller old-growth stands than the owls were known to use. By early 1986, radio telemetry by the Washington Department of Wildlife showed that spotted owl pairs in western Washington range over far larger territories than do owls in Oregon and California. Preliminary figures from the Wildlife Department showed owl pairs using 4,200 acres of old growth. (A more definitive figure published in 1987 was 3,800 acres.) Biologists generally believe the larger home ranges in Washington reflect a smaller prey base there.

Shortly before the draft SEIS was issued, the National Audubon Society published the findings of an independent blue-ribbon Advisory Panel on the Spotted Owl. Appointed by Audubon, the American Ornithologists' Union and the Cooper Ornithological Society, the panel included respected biologists and ecologists from academia and the Forest Service. The panel called protection of 1,500 spotted owl pairs in Washington, Oregon, and California "an absolute minimum" to guarantee the species' long-term survival. The Audubon panel advised the Forest Service to include 4,500 acres of old growth in each habitat area in Washington, 2,500 in Oregon and northwestern California, and 1,400 acres in the Sierra Nevada.

The panel recalled the disappearance of the ivory-billed woodpecker from the southeastern United States after logging wiped out the last of its habitat of old-growth hardwood forests in the 1920s. "Historically, population declines and/or extinctions of North American birds precipitated by human actions have been based on ignorance of one sort or another," the scientists noted. "However, in this case a considered judgment of a federal agency could begin or accelerate an irreversible decline in the Spotted Owl in northern California, Oregon, and Washington."

The shortcomings of the draft SEIS alarmed biologists who had been studying the spotted owl. Not only were spotted owl habitat areas smaller than the home ranges typically used by owls, one-third of all existing SOHAs in Oregon and Washington didn't even have owls. SOHAs were laid out on maps according to a mathematical model intended to distribute habitats widely. In following that model, forest planners left many breeding owl pairs unprotected while setting up SOHAs to protect owls that weren't there.

No allowance was made for catastrophes such as the forest fires that destroyed 100,000 acres of owl habitat in 1987 or the 1980 Mount St. Helens eruption, which leveled 25,000 acres of old-growth forest. The draft SEIS said the national forests' "capability" to support owls would drop from 1,248 pairs to fewer than 600 within half a century. No one knew for certain if that "capability" would be used by real owls or would have value only on paper. No one's assessment was more pessimistic than that of the scientists on the Oregon-Washington Interagency Wildlife Committee. The spotted owl subcommittee advocated a moratorium on further harvesting of spotted owl habitat and called for development of replacement stands for habitat that might be lost to catastrophes: "It is our biological opinion that implementation of any of the other alternatives will lead to the extinction of the northern spotted owl."

When the Forest Service issued its final SEIS in 1988, it contained good news and bad news for the spotted owl. Habitat areas would be of varied size, reflecting the bird's apparent need for a larger foraging area at the northern end of its range. SOHAs would be 2,700 acres on the Olympic National Forest, 2,200 acres in the Washington Cascades, 2,000 acres on the Siuslaw National Forest in the Oregon Coast Range, 1,500 acres in the Oregon Cascades, and 1,000 acres in the Klamath Mountains. Coupled with the graded SOHA sizes were provisions intended to reduce the plan's impact on the timber industry.

In designating SOHAs, Forest Service wildlife biologists were to look first in the wilderness areas and areas already taken out of commercial timber production. Habitat areas would be located on commercial land only if necessary to meet the plan's distribution guidelines. In Washington, the Forest Service touted the increased acreages on the Olympic Peninsula. What the agency didn't mention to the press--and what reporters neglected to ask about--was that the total area in SOHAs was being reduced. Considerably *less* owl habitat would be protected than in the draft SEIS that some scientists said could spell extinction. On the Forest Service's commercially managed forest-land, only one of every six acres of spotted owl habitat would receive protection. The Washington Wildlife Department calculated that the spotted owl population would drop 50 to 77 percent in the national forests of Washington. The Forest Service plan "would likely lead to the extirpation of the northern spotted owl in Washington," wrote Wildlife Director Curt Smitch.

While various government agencies made legalistic arguments in support of or in opposition to the latest spotted owl plan, another drama was playing itself out in California. There, the last wild condor was captured in a desperate attempt to save the declining species. A member of the condor recovery team who had been watching the spotted owl debate offered the judgment that the spotted owl, despite its greater numbers, actually was in greater danger of extinction because of its dependence on a habitat type that was disappearing so rapidly.

Without adequate habitat, a species simply can't survive in the wild. Ann Potter, a young biologist who began tracking spotted owls after working with the endangered peregrine falcon, sometimes was asked why biologists don't do captive breeding of the owl. "With the peregrine falcon," she answered, "captive breeding has had some success because there was some habitat. The species was destroyed but there was a habitat niche for them to return to. There isn't for the spotted owl."

In December 1988, Forest Service Chief Dale Robertson adopted the new spotted owl plan, with one change: SOHAs on the Olympic Peninsula would be increased to 3,000 acres--still smaller than an average pair's home range, still insufficient to give the bird any "wiggle room." Regionwide, the plan reduced the Forest Service's commercial land base, and thus timber sales, by four percent. Only one of every six acres of spotted owl habitat in the commercial land base received protection. Robertson described the plan as striking "a reasonable balance" between the needs of the owl and the needs of the timber industry. The decision was to be effective for five years, at which time it would be reviewed in the light of information then available. Environmentalists promptly filed suit. They scored an important first-round victory, when U.S. District Court Judge William Dwyer enjoined the Forest Service from selling off any large blocks of old growth. His injunction, affecting half of the agency's timber sales in Oregon and Washington, sparked a political crisis. Only in the face of a sudden timber shortage did timber-industry executives show any interest in negotiating with environmental leaders.

The Forest Service's reluctance to tie up much of its timber base in owl habitat areas was understandable. For decades, the agency had been organized primarily to sell timber. The timber program dominated its budget and staff. Timber officers, not biologists or recreation specialists, were promoted to top management positions. Those managers weren't about to sacrifice the timber program to a bird, particularly under an administration that had been talking about doubling or even tripling the national forests' timber sales. It was the job of the Forest Service to balance competing interests. And though its plan for protecting the spotted owl was grossly inadequate, at least the agency was moving in the right direction.

More difficult to explain was the Fish and Wildlife Service's reluctance to declare the spotted owl a threatened or endangered species. Unlike the Forest Service, Fish and Wildlife has but one responsibility: to protect wild animals. When an East Coast environmental group, GreenWorld, petitioned Fish and Wildlife in January 1987 to declare the spotted owl endangered, the only issue properly before the federal agency was to determine whether the species faced extinction or extirpation from a signficant part of its range. Scientists within the agency and from outside offered various models to predict what would happen to the spotted owl population as its habitat was whittled away.

Because of uncertainties about the owl's ability to reproduce and its response to fragmentation of its habitat, the scientists couldn't provide any definitive answers. But their conclusions weren't encouraging. Russell Lande of the University of Chicago, Barry R. Noon of the Forest Service, and Mark L. Shaffer of Fish and Wildlife all predicted extinction or extirpation under Forest Service management plans. Even the industry expert, Mark S. Boyce, was unable to muster much optimism. Boyce criticized the projections in the Forest Service's draft SEIS of population persistence as exaggerating the risk of extinction. However, Boyce was forced to acknowledge a "high risk" of extinction if logging split the spotted owl population into isolated, inbreeding subpopulations.

Because of the political sensitivity of the spotted owl issue, the Fish and Wildlife Service's decision wasn't based strictly on the biological facts. The agency's standard procedure, when considering a species for threatened or endangered status, is for staff scientists to write a status review on the species. The status review contains conclusions and recommendations to be used by policymakers in their decision. Standard procedure was not followed in the case of the spotted owl, however.

Agency biologists were taken out of the decision loop. James F. Gore, the status review team leader, told me later, "We did a status survey on the owl. We were asked at that time not to make any conclusions based on our survey. Then the decision was made in Washington, D.C." Agency managers even deleted from the final version of the status review any reference to the findings of Mark Shaffer, the Fish and Wildlife population biologist who was asked by Gore to review the Forest Service's draft SEIS. Shaffer had predicted extinction, possibly even "more rapidly than habitat destruction." Once again scientists had been consulted and their recommendations for conservative management ignored.

Even while the status review was underway in Portland, Fish and Wildlife officials in Washington were negotiating with the Forest Service. In December 1987, Fish and Wildlife Service Director Frank Dunkle and Forest Service Chief Dale Robertson signed an interagency agreement in which the Forest Service pledged to maintain a "well-distributed population" of spotted owls and to monitor the population to detect any problems. Fish and Wildlife pledged to coordinate habitat protection outside the national forests to supplement the Forest Service's efforts. Two weeks later, the other shoe dropped. Rolf Wallenstrom, then Northwest regional director of Fish and Wildlife, issued his finding that listing the spotted owl "is not warranted at this time."

The General Accounting Office found that ranking Fish and Wildlife officials had compromised their biologists' efforts to assure the "scientific integrity" of the status review. Wallenstrom later confirmed that the decision was based more on politics than on science, telling the press that his superiors had given him little choice in the matter: "There wasn't an atmosphere in Washington to accept a decision to list the owl."

Coming on the heels of Fish and Wildlife's deal with the Forest Service, the decision was hardly shocking. Mark Boyce was distressed, however, when he read the characterizations of his position in the agency's status review. Boyce had been hired by the National Council of the Paper Industry for Air and Stream Improvement to review the Forest Service's analysis of population viability. Fish and Wildlife's status review said Boyce had concluded there was a low probability the spotted owl will go extinct. "I did not conclude that the Spotted Owl enjoys a low probability of extinction," he wrote to Fish and Wildlife's Wallenstrom, "and I would be very disappointed if efforts to preserve the Spotted Owl were in any way thwarted by a misinterpretation of something I wrote."

Environmentalists reacted to the Fish and Wildlife decision by going to court. The case was captioned *Northern Spotted Owl v. Hodel*. In what may have been the most important judicial ruling up to that time on ancient forests, U.S. District Court Judge Thomas S. Zilly found Fish and Wildlife's decision not to list the owl "arbitrary and capricious and contrary to law." The agency's decision flew in the face of expert opinion, he wrote. The agency saw the writing on the wall, announcing in April 1989 its intention to protect the owl as a threatened species. Whether that step would in fact give the owl additional protection would depend on specific measures yet to be adopted.

The Fish and Wildlife Service was in a politically ticklish position when it came to the spotted owl. As part of the Department of the Interior, the agency's director reports to the same Cabinet secretary who oversees the National Park Service and the Bureau of Land Management. Both of those agencies manage forests that provide habitat for the spotted owl. Claiming to be unfettered by the sort of multiple-use laws that govern the Forest Service, BLM managed its 2.4 million acres of land in western Oregon almost exclusively for timber sales. Environmentalists challenged that timber dominance in a lawsuit that sought to force BLM to prepare a supplemental environmental impact statement on the spotted owl. The Ninth Circuit Court of Appeals in 1988 issued an emergency injunction forbidding new timber sales in old-growth forests less than 2.1 miles from a spotted owl nesting area.

The conflicting pressures within the Interior Department could be gleaned from its response to the Forest Service's final SEIS. The Park Service wanted a much stronger spotted owl network in order to protect the owls in Olympic National Park and other parks. "If land uses adjacent to NPS lands alter drastically in the next few years, increasing isolation and fragmentation, we doubt that the spotted owl will survive in the parks," wrote Gerald D. Patton, associate director of the Park Service. The Fish and Wildlife Service, similarly, called for a moratorium on logging of old-growth and mature timber on the Olympic Peninsula, the Oregon Coast Range, and any other place where the owl was in imminent danger. That position, signed in September 1988 by Steve Robinson deputy director of Fish and Wildlife, was notable in light of the agency's decision not to list the owl as threatened or endangered.

BLM's Oregon director, Charles W. Luscher, meanwhile, praised the Forest Service plan for providing "reasonably conservative protection" of the owl while minimizing harm to the timber industry. Luscher raised the possibility of larger habitat areas on the Siuslaw National Forest, where older forests had become highly fragmented. Not surprisingly, the Interior Department's official comments to the Forest Service made no mention of the urgent calls by the Park Service and by Fish and Wildlife for greater protection of the owl. In the Coast Range, Interior suggested only that the Forest Service and BLM "share" land to create habitat areas. Once again, the department's timber-selling interests had won out over wildlife interests.

It isn't only in the federal government that wildlife officials and timber-selling agencies have found themselves in conflict over the spotted owl. In Washington state, the dissension has focused on the western side of the Olympic Peninsula, where the forest meets the sea.

Washington Wildlife Department biologist Dave Leversee has brought his boss, Eric Cummins, to the end of a short spur road bulldozed years ago through this stand of pristine Sitka spruce, hemlock, and candelabra-tipped cedar. The road was built to give loggers access to the valuable timber, but two unanticipated events have kept the trees standing. The timber was sold, but when the recession hit in the early 1980s, the buyer defaulted. The Washington Department of Natural Resources had just resold the timber when a Wildlife Department biologist found a spotted owl nesting area. DNR canceled the sale, to the outrage of many citizens in the nearby town of Forks.

Leversee, a thin blond man, is telling about the first time he saw the male roosting in a hemlock. He returned the next day with another biologist and they "moused" the bird with a gerbil. When the owl flew off with the rodent, the scientists followed it into the woods. They fed it a second rodent and then a third, until they found the nesting area.

The male owl had been feeding his mate and two nestlings. "Now they fly quite well," Leversee says of the young owls.

Leversee is in the middle of the story when Cummins exclaims, "There, Dave!" A spotted owl, its broad wings beating, flies purposefully through the understory. In just a few quick seconds it has disappeared.

"An owl," Leversee responds. "It's good to know one of them is still there."

With some folks in Forks making cracks about going out to "hoot and shoot," the biologists worry about the owls. Relatively safe from predators as long as it stays in the protective environment of the forest, the spotted owl lacks a healthy fear of the unknown. Like a child willing to accept candy from a stranger, the spotted owl is too easily caught by two-legged visitors bearing gifts of mice. (This lack of caution may have been the undoing of two juveniles who were struck by large blunt objects on the Olympic National Forest's Hoodsport Ranger District. A necropsy of the birds established the cause of death but couldn't confirm that the culprits were human.)

No one really expected to find owls in this heavily logged area south of Forks. This isn't a large stand of old growth. The landscape is dominated by clearcuts and young forests, with only about 1,000 acres of virgin timber scattered about. That's not much old growth for a spotted owl pair. The fact that this pair has bred successfully makes it important that their nest be protected. This stand of low-elevation trees may have another value as well. If spotted owls are to move between the central part of Olympic National Park and the park's old-growth coastal strip, then the Hoh-Clearwater

Block--of which this stand is a small part--provides an important habitat link.

The Hoh-Clearwater Block, named after the two principal rivers flowing through the area, contains the last sizeable tracts of old growth owned by the state of Washington. Only during the early 1980s did wildlife biologists and environmentalists closely examine the ecological importance of the 177,000-acre block. So much pristine forest was protected in the adjacent Olympic National Park that few people even considered that any state lands might be needed for wildlife habitat as well. In 1981 Eric Cummins began looking for spotted owls around Kloochman Rock on the edge of the park. By 1983, he and a team of volunteers had located four owl pairs in the Stequaleho Creek and Solleks River drainages. The Department of Natural Resources, meanwhile, was planning two timber sales on upper Stequaleho Creek near the park boundary.

The Wildlife Department asked DNR to withdraw the timber sales. Elimination of owl habitat in the Hoh-Clearwater Block could jeopardize the Forest Service's spotted owl network, Wildlife officials warned. The state-owned land separates the Olympic National Forest's Soleduck Ranger District from the Quinault District by fifteen miles. The Wildlife Department proposed that three 1,000-acre spotted owl habitat areas be established on DNR land to help owls bridge that gap. DNR contended that there was sufficient owl habitat in the higher-elevation old growth on national park lands just east of the Hoh-Clearwater Block. But the Board of Natural Resources, which must approve DNR timber sales, put a moratorium on sales in the proposed SOHAs while a survey was conducted of owls in the park.

During that two-year survey, fewer spotted owl pairs were found in the park than were known to be living on state land in the Solleks-Stequaleho drainage. No owls were found above 2,500 feet in elevation. Olympic National Park's limited ability to support spotted owls is largely a question of topography. From the center of the Olympic Peninsula rises a rugged, snow-capped mountain range. The spectacular Olympic rain forest provides wonderfully rich habitat in the wide valley bottoms and on low-lying hills. But the spotted owl survey showed that upstream, where the valleys become steep and narrow, the number of owls drops off sharply. Along the park's Bogachiel River, said to hold more board feet of timber than any southern state, biologists found fewer spotted owl pairs than on the adjacent lands in the Olympic National Forest.

The owl survey revealed an additional problem. Aggressive barred owls were invading spotted owl territory in the Bogachiel and Queets river drainages. The barred owl, more able than the spotted owl to cope with a landscape fragmented by logging, has expanded its range from the Rocky Mountain region to the coast in recent decades.

Here's how project leader Chuck Sisco summed up the results of the survey sponsored by the Wildlife Department, Department of Natural Resources, and Olympic National Park:

The Olympic population [of spotted owls] may be endangered due to its isolation, low numbers, fragmented habitats, and barriers to population interchange with the Cascade population and within the Peninsula itself. Spotted owls on DNR land in the Clearwater Block occupy a critical position in the distribution of spotted owls on the Olympic Peninsula and should be protected to help provide north-south and east-west links for the westside Olympic Peninsula spotted owl population.

The Wildlife Department recommended that the two Stequaleho Creek spotted owl habitat areas be merged into one large SOHA adjacent to the national park and that five additional SOHAs be created. DNR biologist Deborah Lindley agreed with Wildlife biologists that the spotted owl desperately needed help to survive. She wrote in a letter to a DNR timber manager that the population on the Olympic Peninsula "is on or very close to the edge of a precipitous decline. . . . We are at the point where, in one researcher's words, 'there are no more throw-away owls on the Peninsula.'"

DNR managers rejected the recommendations of Lindley and the Wildlife Department. DNR, managing 1.8 million acres of timberland in trust for such interests as the public schools and universities, was committed to maximizing timber revenues. Any other uses of the land were of secondary importance. In that respect, DNR's management was similar to BLM's. Institutionally, however, the two agencies were very different. Those differences led in the late 1980s to distinct approaches to the old-growth issue. BLM managers, reporting through a typical bureaucratic chain of command, went on with business as usual until the courts said they were operating outside the law. BLM had no legislative obligation to respond to any constituency except the counties that receive the lion's share of timber revenues.

DNR timber sales, by contrast, must be approved by a Board of Natural Resources. The board's chairman, the elected commissioner of public lands, is DNR's chief executive. Members include representatives of the counties, the universities, the public schools, and the governor. Meeting in public, the board's actions are highly visible. The question of how the state would manage its last old-growth stands hit the Board of Natural Resources with a fury in the fall of 1987. That was when an advisory panel of biologists from state agencies, environmental groups, and the timber industry asked that timber sales be deferred for two years in seven "areas of concern." During that period, a comprehensive survey would be made of spotted owls in the Hoh-Clearwater Block.

That recommendation caused a furor on the Natural Resources Board. The areas of concern included 27,000 acres of the state's most valuable timber. DNR was proposing 38 timber sales worth $60 million in those areas. Without those sales, managers said, the trusts would suffer a heavy revenue loss.

In December 1987, DNR staff asked the Board of Natural Resources to authorize six timber sales in the areas of concern. Advocates of the sales spoke of the board's legal obligation to maintain "undivided loyalty" to the trust beneficiaries. Representatives of the Washington Environmental Council and National Audubon Society replied that the board's financial duties didn't override the State Environmental Policy Act. The five-year-old environmental impact statement for DNR's timber management program hadn't considered that the Hoh-Clearwater Block might provide a key link in the spotted owl's habitat network. If the Natural Resources Board ignored the biologists' warnings, they would likely be slapped with a court injunction stopping the timber sales. When the matter came to a vote, the representatives of the governor and the counties blocked the sales.

The reaction from some quarters was swift and angry. Washington's largest newspaper, *The Seattle Times*, editorialized that the board's willingness to give the owl interim protection "short-changes another important species--school kids." The board's action was "an overprotective, priority-poor, out-of-balance cave-in to intense pressures" from the Wildlife Department and environmental groups. *The Times* chided Frank Brouillet, superintendent of public schools, for missing the board meeting and the opportunity to cast the deciding vote in favor of logging the spotted owl habitat.

Brouillet refused to be baited. In a memorandum to Commissioner of Public Lands Brian Boyle, he wrote that protection of the spotted owl and preservation of old-growth forests must be considered along with timber revenues. "I get upset when people say we have to cut trees fast because of schools," Brouillet said later. "We are cutting the trees too fast." To put the revenue question in perspective, Brouillet noted that the timber sales would raise only $18 million toward reducing a $300 million shortfall in the public school construction fund. The timber sales wouldn't solve the funding problem even if the birds were logged out of existence. Brouillet urged that a consensus-building group modeled on the Timber/Fish/Wildlife process be created to provide policy guidance.

Brouillet's courageous stand helped put to rest the clamor for timber sales in defiance of the biologists' advice. (An attempt in the state Legislature to require DNR to sell off spotted owl habitat failed.) In the midst of the hoopla, the Wildlife Commission listed the spotted owl as endangered. Business would not be conducted as usual. With the Natural Resources Board's enthusiastic support, Boyle appointed a 32-member panel to consider how to protect spotted owls and old growth while figuring out how to fulfill DNR's obligation to the trust beneficiaries. The Commission on Old Growth Alternatives included representatives of environmental groups, the timber industry, timber-dependent communities, and trust beneficiaries. During the year that the commission was to deliberate, DNR would be allowed to sell some timber in the areas of concern, but only if the sales met specific criteria intended to prevent further fragmentation of owl habitat. Wildlife Department biologists, under Eric Cummins's supervision, began an intensive, five-month survey for spotted owls in the Hoh-Clearwater Block.

On a summer afternoon, I go with Cummins and Dave Leversee to Owl Mountain, where biologists are studying a male and a female that seem to be foraging on both sides of the national park boundary. We pass Willoughby Ridge, where logging trucks had to be helped up one steep road by a pair of large bulldozers. On the northeast corner of the Hoh-Clearwater Block, DNR clearcuts have sharply etched the rectangular boundary line against the park's old growth. To the south, the Clearwater River has been logged so heavily and rapidly that Cummins shakes his head and says, "This is just nuked up here."

Most of the land on the Hoh-Clearwater Block has been logged during the past two decades, since DNR dropped its nondeclining even-flow policy for the block. (About one-third of the virgin forest is left.) DNR's new policy was to practice sustained yield across the state as a whole, but not within a particular district. DNR timber sales currently are concentrated in the Hoh-Clearwater Block. Logging activity will shift to other areas when the old growth has been liquidated here. DNR is selling 229 million board feet a year annually on this block, but sales will plummet to 39 million in the late 1990s. Then many timber industry jobs may go the way of the spotted owl. DNR's Old Growth Commission is considering whether to put the block back on a sustained-yield basis. But with two-thirds of the land already cut over, even sustained yield would mean a sharp drop in timber sales.

The first three months of the spotted owl survey has turned up five or possibly six owl pairs scattered in various stands of old growth and mature forests. Because some of those sites are separated from any continuous blocks of owl habitat, biologists fear that some of these sites may be reproductive "sinks." Because of their lower reproductive success and higher mortality, animals in a sink are unable to contribute to the survival of the species. One thing is certain: the more fragmented the habitat becomes, the more likely such sites are to become reproductive sinks.

"It's desperate," Cummins says of the spotted owl's population on the peninsula. "If we can maintain what we have right now, that would be a major success. If there were no further cutting, it's hard to say whether this population we have right now would be here in twenty years."

Standing on an inactive log yarding platform, Cummins and Leversee look over the South Fork Hoh River valley while pondering how the spotted owl can best be given a chance of survival. The issue will be considered not only by DNR's Old Growth Commission but also by a separate panel drafting a spotted owl recovery plan for the Wildlife Commission. The 1,000-acre Owl Creek habitat area just south of the Hoh has been only sporadically occupied by spotted owls, none of them breeding. The biologists wonder whether it makes sense to fight for this isolated patch if it means losing stands adjacent to the national park. If DNR instead saves what old growth remains along the South Fork Hoh River, Cummins suggests, the two ownerships could function as a "mega-SOHA."

While Cummins and his colleagues were brainstorming how best to accommodate the owl on state land, federal timber sales on the adjacent Olympic National Forest continued to hack the habitat into pieces. State wildlife officials grew so alarmed over logging in the Soleduck Ranger District early in 1988 that they took the unprecedented step of suing the Forest Service to block a timber sale. It was the first major environmental case to be heard by President Reagan's new appointee to the bench, Thomas S. Zilly. Zilly, who also would hear the important case of *Northern Spotted Owl v. Hodel*, concluded that the Forest Service had failed to take the required "hard look" at the cumulative effects of timber sales. The agency had prepared an environmental assessment but that document hadn't considered how the spotted owl might be affected by many old-growth timber sales in one area. As Sierra Club Legal Defense Fund attorney Todd True had told the judge earlier, "You log a little habitat here and a little habitat there and pretty soon you're talking about real habitat destruction."

Zilly, after agreeing with the plaintiffs on that key point, then reached one of the more curious judicial conclusions of the year. He declined to issue an injunction stopping the contested timber sale because the plaintiffs failed to demonstrate that irreparable harm would come of the timber sale. Even after defining the legal issue as cumulative impact, the judge insisted that the plaintiffs must show harm "ensuing solely" from a single timber sale before he would intervene. The plaintiffs appealed.

If Zilly failed to see the inconsistency of his position, the Forest Service's attorneys didn't. The Forest Service--after ostensibly winning the case in District Court--withdrew the timber sale before the Ninth Circuit Court of Appeals could hear it. Other sales presenting the same legal problem were also withdrawn.

How might the Forest Service and BLM go about protecting the spotted owl if that protection were a top priority of the agencies? Short of a flat-out moratorium on logging of virgin forests, biologists have been trying desperately to get a few basic points across to policymakers. Sites known to have owl pairs should be protected. Habitat areas should be as contiguous as possible. Low-elevation forests--and this generally means commercial timberland--are critical.

Instead, the Forest Service and BLM propose to use wilderness areas and parks as the main population base for the spotted owl. The agencies' spotted owl habitat will function primarily as "stepping stones" for the movement of owls between those population reservoirs. It isn't an approach that satisfies many experts.

Larry Brewer, who headed the Wildlife Department's spotted owl study before joining the faculty of Western Washington University, is worried that the patchwork clearcutting that now covers most of the federal lands may already have spelled the spotted owl's doom. "The problem now is going to be extremely difficult to correct," he observes. "If the U.S. Forest Service had managed the forests initially with some kind of ecological concepts in mind, they would have managed them by drainages. They would have left the waterways, the drainages, timbered so you would have had connecting lines through all of the drainages and through all the subdrainages. And then you could establish some SOHAs down in drainages where there would be connecting timberways."

The importance of habitat corridors has been stated by numerous scientists. Bruce Blanchard, commenting for the Interior Department on the 1986 version of the Forest Service's spotted owl plan, called attention to spotted owls' frequent choice of nest sites near streams and their use of stream corridors for travel. "This suggests," he wrote, "that a more creative alternative be examined: positioning of most SOHAs along riparian zones as bulges in the streamside corridor protection areas."

Perhaps the most credible group of experts on the subject is the spotted owl subcommittee of the Oregon-Washington Interagency Wildlife Committee. The committee's biologists, representing federal and state agencies, industry and the environmental movement, released their recommended management guidelines in April 1988. In areas where owls are most endangered--the Olympic Peninsula, the Coast Range of Washington and northern Oregon, and the Cascades of California--no owl habitat should be logged. Major owl population centers should be protected wherever five to ten owl pairs (three in Washington) are known to make their range in adjacent habitat. Between those "large blocks" should be designated connecting sites of at least 1,600 acres each in California, 2,200 acres in Oregon, and 3,800 acres in Washington. Younger forests within habitat areas should be managed to become spotted owl habitat in time. Habitat areas should be connected by forested corridors at least one-quarter-mile wide. The biologists noted that their proposal "is a compromise and is not the best biological alternative possible." That compromise went far beyond anything seriously considered by the land management agencies.

In the long run, how commercial forests are managed may prove just as important as preservation of existing spotted owl habitat. Where younger forests include key components of old growth--large trees, snags, logs, and a multilayered canopy--those forests can support some spotted owls at a low density. Although students of the owl are virtually unanimous on the necessity for maintaining large blocks of old-growth, many want to see more done to incorporate old-growth characteristics into managed forests. Larry Irwin, a biologist employed by the National Council of the Paper Industry for

Air and Stream Improvement, points to the presence of spotted owls at low densities in parts of logged-over southwestern Washington as evidence that commercial forests can be managed to support some owls.

"I think we have to do that because the old-growth reserve areas and the SOHAs won't last forever even if we attempt to protect that," Irwin says. "The Yellowstone fires and Mount St. Helens show that. It's prudent that we begin to manage the younger forests and create the conditions that can be occupied successfully [by spotted owls] and do it in a shorter time than nature can." Commercial forests will be able to provide this kind of "second-class habitat" for owls, as Irwin calls it, only if management practices are drastically changed. Only a few enlightened foresters are making those kinds of changes. Others have failed to manage the commercial forests to produce new habitat. Except where the courts have stood in their way, managers of the public lands, like private landowners, have logged the best owl habitat out of existence.

The spotted owl is at the center of a drama moving toward tragedy as inexorably as any script penned by Aeschylus. For seventeen years, biologists have played the role of a Greek chorus, warning the Forest Service, BLM, and Washington's DNR that their timber sales were driving the spotted owl toward extinction. Time and again their warnings have been ignored. The bird could well be allowed to fade away in the most public, best understood, most deliberate extinction to take place in modern times.

Officials in the resource agencies spoke of "undivided loyalty" to their trust beneficiaries. Or they spoke of the "reasonable balance" they were striking among competing interests. The survival of a species became just one more interest to be weighed against more pressing political and economic demands. Key appointments made by President Bush during his first months in office seemed to dash any hopes that his administration would show more sensitivity than its predecessor to the plight of the spotted owl and the endangered old-growth ecosystem. Interior Secretary Manuel Lujan, Jr., worried that the timber industry would be devastated if the spotted owl received the kind of protection that the experts felt it needed. "If it is an endangered species, it doesn't need the 3,000 acres per owl or pair of owls. It can include a smaller area," Lujan declared, in disregard of scientific data. As for the Forest Service, the new assistant secretary of agriculture was James Cason. Cason came by way of the Interior Department, where he thwarted efforts to protect the spotted owl under the Endangered Species Act.

The spotted owl had become a convenient scapegoat for the forest-products industry and its employees. Jobs in some mills were disappearing due to automation. Other mills simply closed their doors because they couldn't get enough logs. When the private and public timber supply were both taken into account, the Forest Service's protection plan for the spotted owl accounted for less than a 1.5-percent drop in the Northwest's timber supply.

Around the Olympic National Forest, where timber sales were scaled down more sharply, timber towns were in an uproar. Yet while those towns agitated to restore 60 million to 90 million board feet of timber sales, there was little talk about the *one billion* board feet of raw logs being shipped from Olympic Peninsula ports to Asian nations.

The owl was being sacrificed on the altar of political expediency. The most tragic part of the whole sorry drama was how little would be accomplished by that sacrifice.

11

New Life for the Ancient Forests

"B-b-b-b-buuuurn the dozers!"

The adrenaline pumping through the arteries of the Earth First!ers produces a raucous homage to "monkeywrenching" on the way up the mountain road. Today's action, however, is more in the tradition of Gandhian nonviolent resistance than the sort of environmental sabotage glorified by Edward Abbey in his rollicking novel, *The Monkey-Wrench Gang*. I'm accompanying these *enfants terribles* of the environmental movement to see how they go about shutting down a logging operation.

Suddenly, the song is drowned out by the *budda-budda-budda* of a downshifting logging truck. When a Datsun meets a semi, the car yields the right of way. We stop, back up to a turnout, and let the truck by.

It's carrying large Douglas fir logs, the remains of trees that probably began life before Sir Walter Raleigh set foot in the New World. "*That's* why we're here," observes one of the group, breaking the spell of silence. Once more we head up the one-lane dirt road in the shade of a rich old-growth forest canopy. Talk turns to the realities of jail life and gallows humor about who might become the first Earth First! martyr.

"I'm scared to death," one protester admits.

The Illabot Creek drainage isn't one of the lowest-elevation stands in the Mount Baker-Snoqualmie National Forest. Very little of the valley is below 2,000 feet, and where we're headed is closer to 4,000. But the intact parts of this steep valley are much larger, still, than what's been cut. Moreover, the upper part of Illabot Creek is in the Glacier Peak Wilderness, so the watershed represents a continuous eight-mile-long corridor of old-growth forest. A logging road is being extended into a pristine area north of the creek.

The Earth First!ers don't know yet what they're going to do when they get to the north side of Illabot Creek, where trees are being pulled from a clearcut and where a road is being punched farther across the ridge. They expect forest rangers and perhaps Skagit County sheriff's deputies to be waiting for them. The group had notified the media that there would be an attempt to halt old-growth logging, and at least one newspaper called the Forest Service for comment. During yesterday's hours of training in the techniques of nonviolent civil disobedience, the emphasis was on developing the kind of group rapport that would make quick decision making possible.

228

The group had met at a county park on the Skagit River in the far northwest corner of Washington. The meeting place was a group campsite with a small handmade sign reading "Fourth Corner Trail Club." Friendships were made or rekindled as folks sampled imported beers, wrote post cards to a fellow Earth First!er serving jail time for his actions in Oregon's Siskiyou Mountains, waded in the frigid Skagit, and discussed the bitter split between various Earth First! factions. "I don't know who's politically correct," said George Draffan, an indefatigable Earth First! organizer. "I just know that I have to do what I can to save the forest and save the planet."

Nine young men and women were planning to do civil disobedience, or CD, the next day. Sitting in a circle in the dusk Monday night, they discussed how best to dramatize their opposition to logging of the ancient forest. The protesters spent hours role-playing ways of handling confrontations with loggers and sheriff's deputies, and talking about the risks they would be taking. Felony charges could be filed against them. The logging company could slap individuals with devastating lawsuits to recover damages for lost work time. A peaceful protester could be manhandled or even run over by a truck. They talked about how their arraignments might be handled at the county courthouse in Mount Vernon. Several announced their intention to identify themselves to police as "Doug MacWilliams," the name of the supervisor of the Mount Baker-Snoqualmie National Forest.

They practiced going limp, linking arms to maintain a blockade, and--above all--making group decisions quickly. One of the protesters offered suggestions for defusing anger based on his work with psychologically disturbed adolescents. Another stressed the importance of keeping things peaceful: "If anything goes wrong, we lose."

It was midnight before everyone gathered around the campfire to talk about specific action plans. The discussion was open to the nine who planned to be arrested and to a dozen others in the support group. There was considerable discussion about whether some of the more experienced protesters should try to slip around any police barricade in order to sit in trees or otherwise disrupt road-building and logging. The consensus was that the group should stay together during the action. The meeting ended with everyone circling up and hugging one another. Along with Earth First!'s trademark wolf howls, these so-called "group gropes" would punctuate activities for the next several days.

The sun beats down the morning of the protest as it has every day of this drought-stricken August. Only the woods are cool. The small caravan emerges from the forested shade into a small clearing next to Illabot Creek, where we wait for a KOMO-TV helicopter to arrive. A logging truck roars across the bridge and the driver stops to find out what's going on.

"How do you feel about cutting old growth?" one protester asks him.

"Great!" the driver replies. "I eat spotted owls for breakfast." Off he roars.

The pointed comments lead to a group discussion about whether to talk to loggers during the action. The decision is made not to, because it could inflame their anger and because "our beef isn't with them, it's with the Forest Service." By this time a recon car has returned from the north side of the drainage with the galvanizing news, "There's no closure!" The protesters worry that the Forest Service and sheriff's deputies might arrive at any time to put up a blockade. After all, the loggers with their CBs have seen the group, the KOMO copter has landed, and the Forest Service presumably has been bracing for this action for two days. David Helm suggests that the protesters make their stand as close as possible to the end of the road under construction. That way, they will shut down the operation, a "much more powerful" statement than an action farther from the center of things. Should the group now head toward the end of the road? Everyone gives a thumbs-up sign, and we're off.

Moments later, the Earth First!ers are striding along the unfinished road, heavy chains slung over their shoulders. A few workers on their coffee break watch in wonder as these invaders walk to the end of the road and quickly confer. With the help of the support group, they chain themselves to everything the road builders must move to continue work. One is chained to a large hemlock that's just been felled, others to the clamshell of a huge backhoe, the bumper of a dump truck, and the caterpillar tread of a bulldozer. The signs and banners come out: "Earth First!," "Only You Can Prevent Clearcuts," "Coming Soon: Washington Chainsaw Massacre," and "Stumps Suck."

The protesters are visibly relieved. Whatever happens now, they've succeeded in their basic goal: they've taken a public stand and, at least for a while, stopped the cutting of a road through this ancient forest of western hemlock and Pacific silver fir.

The road builders question the tactics and the intelligence of the "bird watchers," as they call them. Jim Crabtree, the foreman and backhoe operator, has radioed his bosses at Miller Shingle in Granite Falls and been told to stop work and leave the handling of the protesters up to legal authorities. Crabtree and the two truck drivers keep a respectful distance from the Earth First!ers as the hours pass. The road builders observe that the protesters had to use roads like this one to get here. And if the protesters live in wood houses and use paper products, they should understand the need for logging.

Besides, the grizzled, chain-smoking Crabtree says, "Go take a look at the trees. They're all rotten, 50 percent of them. They'll all be like that tree lying right there on the ground." A couple hours later Norm Strand, owner of a timber company logging a nearby clearcut and the mayor of Concrete, arrives. He wonders why anyone would want to save this forest. Forty percent of the timber is only good for pulp, he says. Then, gesturing toward the protesters, he comments, "I don't think there's any of them I would want to have vote for me. I don't see any I would want in the Log Cabin Tavern."

Eventually, a green Forest Service pickup pulls up and a protester shouts,

An Earth Firster chains himself to an old-growth hemlock log in an effort to stop construction of a logging road on Illabot Creek in northwestern Washington.

"Here come the Freddies!" The protesters break into song: "We don't believe you any more." Warned that the Forest Service will not tolerate their hindering the road building, Mitch Friedman responds, "Our intention is not to hinder the operation. Our intention is to save the forest. If that means hindering the operation, that's an unfortunate side effect."

As the afternoon drags on, it gradually becomes apparent that there will be no arrests. George Draffan reports by radio from the county park in Rockport that a sheriff's deputy has been in and out of the park, using the telephone in a fruitless attempt to locate a pair of chain cutters. The forest rangers tell the press that the forest supervisor hasn't authorized the road closure that would make arrests possible. Convinced that they've stopped the road building for today, the protesters unchain themselves, circle up, and make a quick decision: they'll return to their campsite and try again tomorrow to disrupt logging operations.

Back at the park, the Earth First!ers do another group grope while discussing their options. They decide they'll try to block the main Illabot Creek road tomorrow, preventing the transport of logs from the new road and the clearcut. They call the Skagit County sheriff's office to report the intended crime and urge their own arrest. The next day, the protesters sit in the road in front of a logging truck. They keep the road closed for the rest of the day. Once again, the Forest Service declines to initiate arrests.

The Forest Service strategy serves two purposes: one, outlasting a protest that will run out of steam of its own accord, and, two, denying free publicity to Earth First! by refusing to make arrests. The strategy works on the first count: by the end of the week, civil disobedience has ended. On the second, the Forest Service strategy has failed. Earth First!ers, masterful manipulators of the media, once again have gotten big play with small numbers of people and this time without even having to go to jail. They've staged the sort of drama that makes the news business go round: brave young men and women risking their freedom, even their lives, to save the forest from rape at the hands of the Forest Service. It's the kind of publicity that money can't buy.

Before ancient forests became a top priority for the environmental lobby, no group did more than Earth First! to awaken the public to the destruction of the old-growth ecosystem. Earth First!ers submitted to arrest to block a logging road in the North Kalmiopsis area of the Siskiyous in 1983. They perched in the giant Douglas firs of Millennium Grove two years later, moving more than a few Oregonians to tears as the forest was cut down around them. Much like the Vietnam-era protesters who sat in front of weapons trains or poured blood on draft files, Earth First! brought the old-growth issue into public focus by reducing a nettlesome public-policy debate to a stark moral choice.

Staking out the high moral ground has its practical limits, however. The actions of a small, anarchic group shouting, "No compromise in defense of Mother Earth!" dramatize the issues but don't suggest how the crisis of the Northwest forests will be resolved. When monkeywrenchers throw a pie in a state forester's face, sabotage logging equipment, or endanger sawmill workers' lives by spiking trees, public debate inevitably shifts from the forests to those questionable tactics themselves. That point has not been entirely lost on Dave Foreman, the one-time Wilderness Society lobbyist who spearheaded the Earth First! movement because of his disenchantment with Capitol Hill-style environmental compromises. "There are rational solutions," he told one audience, "but others should find them."

Mainstream environmental activists, more inclined to use the courts, lobby Congress, and make compromises, can point to more practical results. Activist Susan Saul and others in the Gifford Pinchot Task Force persuaded Gifford Pinchot National Forest officials to defer timber sales in most of the pristine 58,000-acre Dark Divide until a new forest plan was adopted.

Tom Campion, a Pilchuck Audubon activist who lives in the logging town of Snohomish, Washington, has spent years calling in spotted owls and monitoring timber-harvesting plans on the Mount Baker-Snoqualmie National Forest. After calling in a spotted owl on the North Fork Snoqualmie River, he persuaded the Forest Service to move a timber sale to a higher elevation. In the Canyon Creek watershed, he turned his anger over the 166-acre Olo Mountain clearcut into a different kind of action. Campion signed an affidavit used by the Sierra Club Legal Defense Fund to take the Forest Service to court. The judge halted the clearcutting, and the Forest Service agreed to accept appeals of timber resales.

Sometimes a patch of forest is saved in a more amicable way. Such was the case with Noisy Creek, a steep, glacier-fed waterway just below the Noisy-Diobsud Wilderness. Environmentalists became concerned about the spectacular old-growth stand when Scott Paper sought a state permit to log 160 acres of its 629-acre inholding in the Mount Baker-Snoqualmie National Forest. Logs would be floated across Baker Lake by barge. Company officials said they planned to log the entire property over several years.

Frank Ancock of North Cascades Audubon Society was determined to stop the logging. The Scott Paper property was the only private property in a pristine roadless area. It was excluded from the Noisy-Diobsud Wilderness only because it was in private hands. All of Noisy Creek was in the wilderness except its mouth, which ran through Scott Paper land. The Scott property also was the only obstacle to rebuilding a lakeside forest trail.

Activists organized Friends of Noisy Creek. They fought the company to save the grove of huge cedars and Dougas firs. The Whatcom County Council denied Scott a shoreline management permit for the clearcut. That gave Friends of Noisy Creek time to try seek final resolution. Ancock and others in the group worked simultaneously on publicizing the endangered property while seeking help from members of Washington's congressional delegation.

Representative Norm Dicks, a member of the Appropriations Committee, took up the cause of Noisy Creek, securing a $2.5 million appropriation from the Land and Water Conservation Fund for the Forest Service to buy the site. Federal ownership would preserve a highly visible roadless area and allow reconstruction of the East Lake Trail.

Important as they are, actions to save individual old growth stands won't resolve the broader crisis of the public lands in western Washington and western Oregon. The central issue is one of how the large tracts of virgin forests will be managed. How much will be available for timber sales, and how much will be set aside for protection of the old-growth ecosystem? Until that basic issue of land allocation is resolved, Earth First!ers will continue to sit in front of logging trucks, truckers will form protest caravans, judges will have to review timber sales, and elected officials will hear urgent pleas about the fate of watersheds they've never heard of.

During the 1980s, the tide began to rise in favor of environmentalists. The Forest Service made clear its intention to slow timber sales on the west side of the Cascades. The national forests' capacity for sustained-yield timber sales dropped in response to the Oregon and Washington wilderness acts of 1984 and the requirements of the National Forest Management Act of 1976. The forest-products industry was placed in the defensive position of trying to show why timber sales shouldn't be cut back as much as the Forest Service has proposed. Above all, the industry had to contend with that feathered thorn in its side, the spotted owl. Court orders in the late 1980s sharply curtailed federal sales of old-growth timber.

"Where does it stop?" asked Bob Dick of the Washington Forest Protection Association. "With RARE I [Roadless Area Review and Evaluation, Round I], they set aside these lands. We won some and we lost some. With RARE II they set aside more wilderness areas in the state and a volcanic monument. We're getting into the planning process now and we hear the same arguments we've been hearing for fifteen years: we've got to have more, we've got to have more. And our people are saying, 'Good grief, how much more is there?'"

In victory lay peril for the defenders of the ancient forests. If the federal courts were to give a resounding victory to environmentalists and the spotted owl, a full-blown crisis would result. With billions of dollars of timber in jeopardy, industry would take its case to Congress. If necessary, the industry could be expected to throw its resources into trying to rewrite the laws protecting endangered species. As the 1980s came to a close, the environmental movement and the timber industry were bracing for a titanic struggle.

Can the old-growth issue be resolved short of all-out warfare on Capitol Hill? Are there outcomes that would meet the needs of the spotted owl and the forest-products industry? Certainly, there's no way to fully satisfy industry and environmentalists. There just isn't enough old growth left. Painful choices will have to be made.

Sooner or later, there must be a resolution of some kind. It can be imposed by a Congress that would rather not wade into the quagmire, or it can be negotiated by interested parties that see a greater cost in continued fighting than in accommodation. Hope for a negotiated settlement grew in 1989, with environmentalists' victories in court and before the Fish and Wildlife Service. There's no way to predict or dictate the terms of a settlement. But at least some of the grounds for discussion can be identified. If the needs of industry and the environment are to be met, the private timberlands must be considered along with public lands. Here are some suggestions for restoring peace to the Northwest forests.

Let's make our exports work for everyone. After years of record trade deficits, it's refreshing to see currency flowing to the United States for forest products. It's also encouraging that exports of lumber and other finished products rose substantially during the 1980s. But the fact remains that, like a Third World country, most of the Northwest's forest-products exports are raw materials that are converted into finished products by mill workers in Japan, China, and Korea. Independent mills are being squeezed between the high prices paid by Japanese trading companies for top-grade old-growth logs and the more modest domestic prices the mills receive for wholesale lumber. If those mills could afford to buy more logs from state and private land, they would be that much less dependent on federal sales of old-growth timber.

There may be a solution to the export issue more effective and less Draconian than the ban that independent mills would like to see placed on the export of raw logs from state lands. The federal government could tax log exports while leaving the export of finished products untaxed. A modest duty would create an additional incentive for foreign buyers to buy American lumber and plywood rather than American logs. But its greatest value could be the revenues it produced. These revenues could be used to purchase carefully selected timberlands for addition to the national forests. By enlarging the national forests' base of commercial timberlands, more wood would be made available to the mills that are dependent on federal timber sales. Even if much of this land had very young trees or no trees at all, it would produce a future timber supply that would boost the allowable sale quantity immediately.

Land would be bought only from willing sellers--primarily land intermingled with or adjacent to national forests. Old-growth stands could be bought for their habitat value. Forest stands of less ecological value elsewhere could then be freed for timber sales. If a 5-percent export duty had been in effect in 1987, $28 million would have been raised for this old growth/independent mill preservation fund. With timberland selling for as little as

$100 to $200 an acre, that money could buy a lot of land. Even if the Forest Service had to pay $1,000 per acre in a hotter market, one year's revenues could buy 28,000 acres.

Congress could delegate the authority to spend export tax revenues to a board with representatives of industry, environmental groups, and the Forest Service. In addition to land purchases, the board could be authorized to provide assistance to any loggers or mill workers displaced by reductions in old-growth timber sales.

Let's get more private land into timber production. The lands owned by the forest-products industry are the most intensively managed, most productive timberlands in the Northwest. Yet because of "the gap"--the lack of second-growth trees reaching harvestable age--timber harvests on industrial lands will drop in the 1990s. So will timber sales on the national forests.

Where significant productivity gains are possible is on the nonindustrial private forests. These four million acres in western Washington and western Oregon are grossly understocked with timber. One-tenth of these lands capable of producing commercial softwoods in Washington doesn't even meet the state's definition of being stocked with timber. Yet acre for acre, this Third Forest can grow wood faster than the national forests can. The nonindustrial lands generally enjoy the lower elevations and gentler topography more conducive to tree growth. Western Oregon's nonindustrial forests are believed capable of a fourfold increase in timber production.

The nonindustrial private forests could play an important role in seeing the forest-products industry through some trying times. Economic consultant Philip L. Tedder told a forestry conference in Salem, Oregon, "As the forest- products industry scrambles to locate timber supplies to replace their inevitable decline in harvesting on their own lands, Forest Service harvest levels decline 20 to 30 percent on individual forests, Southern forest softwood inventory levels decline, and Canadian harvests do not increase as planned, the nonindustrial private forestland owner will be there with the timber and the price paid will be premium."

If the nonindustrial private forests are to be a continuing and substantial source of timber, their management must be improved. Acreage in the zero- to 40-year age classes is in disturbingly short supply. Those are the forests that will be important in the next century. Surveys of forest owners have shown considerable interest in more intensive tree farming. But federal programs to give small landowners technical advice and to share costs were emasculated during the Reagan years.

The laws of supply and demand will help rectify the situation somewhat. If timber shortages cause prices to rise, it's a good bet that nonindustrial owners will make greater investments in their forests. The problem is that the front-end investments in reforestation don't pay off for another four to six decades. If we're serious about reducing the timber industry's dependence on the national forests' old growth, increased investment in nonindustrial private forests is critical.

Let's do something about the loss of private timberland. Government cannot and should not tell property owners that they can only grow trees on their land. But state and local government seem not to be heeding warnings from the Washington Department of Natural Resources' that half a million acres of private and state forestlands will be taken out of forestry between now and the year 2040. Most of those withdrawals will be around fast-growing urban areas such as Seattle and Everett in Washington, and Portland and Eugene in Oregon. Tax breaks currently available aren't slowing down the loss of forestlands. Too often, local government inadvertently seals the fate of forests by forcing their owners to share in the cost of utilities serving the very developments that push them out of business. The financial rewards of subdividing are so much greater than forestry that all but the most committed tree growers sell out. Even Weyerhaeuser, "The Tree Growing Company," has no qualms about building condos and office parks where valuable trees could still be grown. With their existing arsenal of weapons, state and county officials are helpless to stanch the hemorrhaging of the timber base.

We needn't remain helpless. King County, Washington, showed in the early 1980s that it could, through incentives rather than penalties, stop the conversion of farmlands to suburban development. Voters approved $50 million in general-obligation bonds to buy development rights from landowners whose farms were threatened by suburban growth. The program, though costly, was a rousing success. There's no reason a similar approach couldn't be used to save the endangered forests on Oregon and Washington's urban fringes. Not only would this help maintain the supply of second-growth timber, it would enhance the quality of life for county residents. As one concerned forester says of the financial and environmental costs of sprawling development, "Even a mismanaged forest is better than a parking lot."

Let's put our growing knowledge about the forest ecosystem to work on commercial forests. Today's tree farms are as different from natural forests as night is from day. By the time a timber stand is harvested, usually after forty-five or sixty-five years in the Douglas fir region, it has developed little biological diversity. It lacks the wealth of large logs, snags, big trees, mushrooms, lichens, and mosses that supports so many amphibians, mammals, and birds in an old-growth forest. A "managed forest" isn't really a forest at all. It's a stand of nearly identical trees--identical in species, age, size and, increasingly, in their genetic makeup.

Tree farms aren't run, of course, for the benefit of wildlife or some hard-to-define ecosystem. But the forest decline being experienced in Europe and the southeastern United States raise questions about the sustainability of forests managed as an even-aged monoculture. Enlightened self-interest demands that timberland owners consider whether the long-term productivity of the land is being sacrificed for short-term financial gains. Research in the old-growth forests of Oregon has shown that trees, small mammals, fungi, and bacteria all work together for their mutual benefit. Restoring and maintaining those relationships will help preserve the productivity of commercial forests.

Roughly five-sixths of the forestland of western Washington and western Oregon has been transformed from a natural ecosystem to a landscape controlled by humans. The other one-sixth, generally on the lower slopes of some of North America's most rugged mountains, can't sustain wildlife in safe numbers without some help from the lowland forests. How we manage our extensive commercial forestlands may be as important ecologically as preserving more old growth. If we are to follow the advice of forest ecologists to leave more large trees untouched from one commercial rotation to the next, the Forest Service will have to lead the way on its five million acres of commercial forestland in the Douglas fir region.

Let's not fragment the last intact tracts of ancient forest. From the window of a single-engine plane flying at low altitude over the Mount Baker-Snoqualmie National Forest, the land below looks like a patchwork quilt. In every major drainage, with the exception of a few roadless areas, the scene is the same. The background is the deep green of tall trees, the rough and ragged look of old growth. The foreground is what's cut out of that wooded background. Throughout each valley are dozens of clearcuts and dozens of uniform stands of young trees. None of these cuts is so large that Aunt Mabel is likely to write back to her neighbors in Oshkosh exclaiming that a whole mountainside is being stripped at one crack. Taken together, though, these clearcuts add up to the slow death of an ecosystem. By the time young regrowth becomes the background and a few isolated old-growth stands the foreground--as already has happened on much national forest land--the ecological integrity of the forest is long gone.

Crossing the border into Canada, the artificial landscape is entirely different. In contrast to the American pattern of odd-shaped cuts scattered randomly over the drainage, there is no question that the loggers of these Crown lands mean business. A road is cut straight through the middle of a valley. Large rectangular clearcuts, one starting where the last ended, march toward the upper end of the valley. From there, the road curves uphill and straight back down the valley, a second tier of clearcuts taking out another third of the forested hillside. There is little regrowth in these valleys; the clearcuts are too new. The ancient forests are going fast.

At first glance, the American logging method seems more environmentally sensitive. Smaller clearcuts mean big game animals can browse close to cover. The slower pace of cutting means less severe damage to water quality. But in the long run, the American way will prove no less damaging to the old-growth ecosystem than the Canadian method. Once-continuous stretches of untouched forest are being reduced to islands in a sea of clearcuts. The remaining old-growth stands are too small to produce the distinctive micro-climate found in virgin forests. Trees once protected by tall neighbors are now vulnerable to blowdown. The salvage of windthrown trees has become a big business in the national forests.

Fragmentation of the ancient forests is proving disastrous for the spotted owl and other species that depend on the old-growth habitat. If public land managers are to continue selling virgin timber, there are two ways of doing that. Sales can be spread out as much as possible--thus fragmenting the old-growth forests everywhere--or they can be concentrated in some drainages while providing intact habitat corridors in others. The second alternative makes more sense. The roadless areas provide far more protection to the old-growth ecosystem than do all of the Forest Service's and BLM's spotted owl habitat areas. Even at the risk of a divisive RARE III battle, the roadless areas must not be opened to logging until their contribution to the ecosystem has been thoroughly analyzed.

Let's give new life to the old growth. There are some winners and plenty of losers in federal and state governments' old-growth timber sales. Independent sawmills benefit from federal sales of high-value timber in a restricted marketplace. Small landowners lose because they can't grow and sell timber as cheaply as the national forests can sell the old growth put there by God. Timber buyers go after what's cheapest and easiest to get. These days that's the old growth on federal land. We're caught in a vicious circle. It is argued that private landowners can't take up the slack if the national forests reduce timber sales. Yet it is precisely the bargain-basement sales of public timber that discourage private owners from managing their forests more intensively.

Tourists and backcountry recreation enthusiasts spend billions of dollars each year to hunt, fish, hike, and camp in a pristine environment. They and the businesses they support also lose out from old-growth timber sales. Tourists aren't drawn to clearcuts and second-growth forests. The billion-dollar salmon and steelhead fisheries suffer as more old-growth timber is cut on unstable slopes, choking spawning beds with sediment. The clean, even water flows produced by ancient forests allow citizens to avoid spending tens of millions of dollars on municipal filtration systems.

Then there's the forest itself--what's left of the marvelous old-growth ecosystem. Biologists finally have put the notion that old growth is a "bio-logical desert" where it belongs: in the trash heap, along with a recently retired president's unfortunate comment that trees cause pollution. Yet we haven't reached a political consensus that the destruction of this continent's last virgin forests must be slowed down, much less stopped. First the Forest Service chose the spotted owl as an "indicator species" for the health of the old-growth ecosystem. Then, when the agency's management plans guaranteed the decline, if not the utter ruin, of the indicator species, habitat areas were shuffled around in a way that did little to improve the spotted owl's prospects for survival.

Peter Morrison's inventory of national forest land meeting ecological criteria for old growth, showed that only a fraction of the amount claimed by the Forest Service actually exists. Little true old growth remains, and precious little at the lowest elevations. The burden should not be on ecologists and environmentalists to show why an ancient forest should be saved. The burden should be on those who would cut it to show why the forest is not ecologically critical. Logging has fragmented the Pacific coast forests so severely that many old-growth groves have lost most of their ecological value and thus can be logged with little further damage. Logging of late-succession forests must be halted altogether where the spotted owl is in greatest jeopardy: the Olympic Peninsula, the Oregon Coast Range, and Washington's Cedar and Green river watersheds.

While focusing on the true old growth, it's important to remember the importance of mature forests that meet some but not all of the criteria for determining old growth. For many species, these forests provide just as much habitat as old growth. These woods cover a greater part of the landscape than old growth, and they are the replacement stands for old growth that will succumb to the insults of nature or mankind. Replacement stands are critical. All of the remaining natural forests have become rare; old growth is the rarest, and most valuable, of the rare.

Let's develop a vision broad enough to restore peace to the forests. The Forest Service, careful to steer a safe course between the "extremes" of environmentalists and the timber industry, has sent its boat down the middle of the river--straight toward the waterfall. The agency isn't about to capitulate to the demands of the environmental movement. And it can't legally go along with the timber industry's demands. So the Forest Service has struck what its managers perceive as a "reasonable balance" among competing interests. That "balance" has repeatedly fallen short of the law, as interpreted by the courts. Inadequacies in the Forest Service's spotted owl plan are likely to bring the timber program to a crashing halt. If environmentalists win in the higher courts, the conflict will move to a different venue: the halls and cloakrooms of Congress. That process is a crapshoot for both sides. And although Congress is an excellent forum for making deals, those deals stop at the limits of federal authority--too short for a settlement that could meet the bottom-line needs of an endangered industry and an endangered ecosystem.

Even as the opposing sides gathered their forces for the final conflict, promising avenues for compromise remained unexplored. Departure from the even-flow version of sustained yield could provide temporary relief to independent sawmills until second- and third-growth industrial forests reached harvestable age. Reduced federal timber sales could be offset by voluntary or mandatory reductions in raw log exports, by increased timber production on nonindustrial private forests, and by policies aimed at discouraging conversion of forests to other uses. Yet even as state timber exports were put on the bargaining table, the industrial landowners continued to ship raw logs overseas. The Northwest lacked an industrial policy that assumed jobs mattered.

The single step that could accommodate the forest ecosystem and the timber industry was expansion of the public lands. Federal and state government could purchase productive second-growth timberland from willing sellers, reserving its timber for domestic processors. The new public lands would be operated as tree farms, the remaining virgin forests as reservoirs for biological diversity. Except for a dwindling number of specialty mills, the industry could be freed from its dependence on old growth. Washington Public Lands Commissioner Brian Boyle's proposal that his state buy 300,000 acres of timberland marked an important move in this direction. Unfortunately, other public officials and advocacy groups were slow to recognize expansion of the public lands as a solution.

The single step that could accommodate the forest ecosystem and the timber industry was expansion of the public lands. Federal and state government could purchase productive second-growth timberland from willing sellers, preserving its timber for domestic processors. The new public lands would be operated as tree farms, the remaining virgin forests as reservoirs for biological diversity. Except for a dwindling number of specialty mills, the industry could be freed from its dependence on old growth. Washington Public Lands Commissioner Brian Boyle's proposal that his state buy 300,000 acres of timberland marked an important move in this direction. Unfortunately, other public officials and advocacy groups were slow to recognize expansion of the public lands as a solution.

The old-growth dilemma had become a test of our ability to solve a tough natural-resource problem creatively. "Ultimately, the handling of old growth in the Pacific Northwest is apt to emerge as the most important natural resource issue in the last half of the twentieth century," noted wildlife biologist Jack Ward Thomas. "How the issue plays itself out will determine how natural resource conflicts are played out in the twenty-first century."

But by the end of the 1980s, the political tide seemed to be turning. Ancient forests had become such a prominent issue that the rank and file of the environmental movement was ready as never before to fight for forests in the aggregate. Hundreds of activists from around the world flocked to Seattle for a 1988 conference to map out a strategy to fight the destruction of rain forests in the Northwest and elsewhere. The top leadership of the nation's largest environmental groups met with West Coast activists to create an Ancient Forest Alliance. National Audubon Society Vice President Brock Evans proposed establishment of an Ancient Forest Preserve System to protect old-growth forests wherever they exist. Forests had risen from an after-thought in specific wilderness bills to the top of the environmental movement's agenda.

Local and regional conservationists had awakened the national leadership to the fact that time was running out on the ancient forests. Intractable though the conflict seemed, the warring parties could take heart from several consensus-building precedents in Washington and other states. Decades of legal--and sometimes armed--struggle between white and Indian fishermen had been settled through direct state-tribal negotiations. Environmentalists and the timber industry buried the hatchet over state forest practices rules in the Timber/Fish/Wildlife process. And finally, a plethora of interests was attempting to come to terms over the management of state-owned old growth on the Olympic Peninsula. At the national forest level, there was no reason the planning processes couldn't be modified to give competing interest groups a crack at hammering out some kind of compromise.

Bringing parties to the old-growth issue to the table would take strong political leadership, leadership that did not seem to be on the horizon. Certainly the Forest Service showed little interest in opening up its planning process to the difficult problems associated with building consensus. To be sure, it wasn't clear whether the national forest planning process was the best place to seek accommodation since state and private lands would be key elements of a comprehensive settlement. Industry and environmental interests were reluctant to bargain as long as each saw a chance of winning more through continued struggle.

Court orders putting the brakes on timber sales in spotted owl habitat gave sawmill owners a motivation to negotiate. Environmentalists saw benefits in bargaining, too, because the increasingly conservative federal judiciary could at any time give the Forest Service the green light to resume logging 48,000 acres of natural forests each year. Politically, time seemed to be on the side of environmentalists. The winds of public opinion were blowing in a new direction: the ancient forests must be saved. If the timber industry chose to be part of the solution, it might win the support of environmentalists in obtaining help in solving its timber supply problems. If not, industry might have to be left behind.

A tree stands in what was, until a short time ago, an old-growth rain forest. It's a large tree, 178 feet high, 19 feet in diameter. In fact, it's the largest tree of its kind, western red cedar. Small hemlocks, salal, and huckleberry grow from crevices in its bark. It stands alone now, on a hillside covered with saplings planted in 1976. If trees silently weep, this one must be in deep mourning.

Randal O'Toole tells of a discussion that turned to that tree. Washington Department of Natural Resources foresters were talking about large trees during a discussion with environmentalists in Spokane when one of the DNR people mentioned the cedar. The enormous tree was first noticed by a logger

The forest is gone, but the world's largest western red cedar still stands. The nineteen-foot-diameter giant is the only reminder of the magnificent forest that once stood on this Olympic Peninsula hillside.

working for the timber purchaser, ITT Rayonier, on an Olympic Peninsula clearcut. The company spared the tree but cut another nearby that was slightly smaller.

The environmentalists were aghast that the world's largest cedar was left without any protection from the wind. Weren't the foresters worried the tree would blow down? "Oh, no," answered one. "Cedar resists decay. If it blows over, we'll be able to salvage it all."

This is big-tree country. A few miles southeast of the largest cedar, the world's largest Douglas fir can be found on the Queets River. South of that, at Lake Quinault, is a Sitka spruce that's tied with an Oregon tree for another world title.

Sheer size isn't the point, of course. From almost any perspective, old growth is a treasure. Ecologists can define something of what's special about an old-growth forest. Practitioners of the Salish Indians' ancient *seyowen* rituals can define other special qualities. Where loggers find board feet of wood fiber, native dancers encounter their spirit power. The day after visiting the big cedar on DNR land, I hike through the Bogachiel River rain forest in Olympic National Park. Chain saws were allowed into the park to take out some large spruce during World War II. Only a small part of the park was cut to feed the war machine, so most of the upper valley remains classic, untouched old growth. Large spruce, cedar, Douglas fir, even a few giant cottonwood, abound. Thick layers of moss and lichen cover bigleaf maples. This profusion of life would satisfy a Hollywood director's image of a jungle. During my brief visit, I encounter a great blue heron, grouse, hare, frog, and two varieties of snake. Deer and elk droppings litter the trail, and during this unusually dry rain-forest summer, a tiny crayfish has been left stranded in the still-wet bootprint left by a hiker. There is no creek or pond within view.

This valley, like the national forest lands just to the north and west and the state lands to the south, is rich in spotted owls. Marbled murrelets have been found up the hill on Rugged Ridge. Humans, too, thrive here. Four backpackers, skinny-dipping in the Bogachiel, tell me they've come here to avoid the crowds on the better-known Hoh River. A backpacking Park Service volunteer calls the valley "one of the best-kept secrets in the park" and says he would almost be willing to pay for the privilege of doing his job.

I'm reminded of the fragility of the forest as soon as I leave the park. Towers are set up to yard logs on an enormous clearcut on private land. Highway 101 winds through clearcuts and second-growth forests on land that once grew whole valleys full of trees like the last big cedar. I catch glimpses of old growth as my car passes through the coastal strip of the national park. The strip is, on average, only a mile or two wide. From there, it's another 170 miles to Seattle, and there isn't an old-growth tree to be seen. I marvel at how quickly the landscape has changed in just one century. Where there once was mile upon mile of proud, magnificent forest, there are now second-growth trees with only brown duff underneath, pastures, homes, taverns, schools, and ball fields. Even in present-day Seattle, mighty cedars like those on the Peninsula once abounded.

Joni Mitchell's song keeps running through my mind:

> Don't it always seem to go
> That you don't know what you've got till it's gone?
> They paved paradise
> And put up a parking lot.

> They took all the trees
> And put 'em in a tree museum,
> And they charged all the people
> A dollar and a half just to see 'em.

Olympic National Park is a museum, one of the grandest on the planet. It protects a few trees from the logging that's still changing the face of the Olympics, the Cascades, and the Oregon Coast Range. It's not that some old growth won't be left in the national forests. Some will. But for the most part, the kind of old growth found in the Bogachiel Valley is on the auction block.

Isn't it ironic that the destruction of the old growth is so often justified by pointing out the "decadence" of the dying giants? As if we do a service to the rotting trees and the animals that live in the trees to cut them down and haul them away.

One of my acquaintances who has a cabin near Darrington keeps a close eye on what's happening in the woods. The more clearcuts he sees, the more he rails at "the bastards"--referring collectively, I suppose, to the Forest Service and its timber buyers. Given his love of backpacking in the wilderness, his anger is understandable. But the timber industry also is an important part of the Northwest economy and culture. The question is how much of the last ancient forests--if any--should be sold off to support this industry. Increasingly, old-growth timber sales help one industry at the expense of another. It's time to drop the idea that the economy benefits from every timber sale.

The timber industry asks: can we afford not to sell off all the old growth outside the parks and wilderness areas? More to the point, I wonder, can we afford to sell off a national treasure so cheaply?

When my older son was five years old, he told a friend's mother one day that she shouldn't sell books because trees had to be cut down to make the paper.

"Oh, they can grow new trees," she replied.

"But they're not the same," Daniel shot back. Some of the political and economic complexities eluded him, as they do all of us. But he had learned an ecological fact that we must not lose sight of: once cut down, an old-growth forest is effectively gone for good. A century from now, the national forests and parks may offer little more than trees in a museum. Or they can support vibrant, changing, *natural* ecosystems. The last primeval forests offer us an opportunity. We can avoid the mistakes that we made on the rest of this continent and that are still being made in the tropical rain forests.

The westward sweep of European civilization on this continent has laid low every forest in its path. The once-grand woodlands of the East, the South, the Midwest are gone, with only the smallest of remnants here and there. As little as 2 percent remains of North America's primeval forests. These forests are primarily in the Rockies and the coastal range from northern California to southeast Alaska. The Douglas fir forests of Oregon and Washington, together with the cedars of southern British Columbia and California's redwoods and sequoias, are the jewels of North American forests. They are in fact, the most spectacular conifer forests on the face of the earth.

The old-growth forests of the Pacific Northwest have become even more valuable in this era of global deforestation. Two thousand years ago, the shores of the Mediterranean Sea were largely transformed from forestland into desert. From the Oklahoma Dust Bowl of the 1930s to poverty-stricken sub-Saharan Africa of the 1980s, humankind's reckless removal of the planet's vegetative cover and the land's topsoil is bearing disastrous results. The world's biological diversity is becoming impoverished, and deforestation is accelerating the buildup of atmospheric carbon. The greenhouse effect resulting from that buildup could conceivably reduce much of the Northwest forest to dry, California-style chaparral. The best defense that forestry can offer against climatic change is to leave intact the genetic resilience and carbon-absorbing mechanisms of the old-growth forests. Every tree is a water pump that transpires moisture into the atmosphere.

The ancient forests are one piece of the larger puzzle that makes life on this planet possible. Years of debate and research lie ahead as we ponder how that piece fits into the puzzle. There's a simpler truth as well, that can be divined in the sheer majesty of the forest. This truth can be seen in the green fire of a dying wolf's eyes or heard in the cry of the seaward-flying murrelet. It's a truth as ancient as the forest, older than *Genesis*: "And God saw every thing that he had made, and, behold, it was very good."

Selected Bibliography

This reading list suggests a few excellent sources of information about the old-growth forests of the Pacific Northwest. A more extensive list can be drawn from Notes.

Andrews, Ralph W. *This Was Logging*. Seattle: Superior Publishing Company, 1954. Andrews' text is secondary in this remarkable collection of Darius Kinsey's photos of logging's early days in the Northwest. Andrews' *Redwood Classic* (Superior Publishing, 1958) draws from several photo collections.

Cox, Thomas R. *et al. This Well-Wooded Land: Americans and Their Forests from Colonial Times to the Present*. Lincoln: University of Nebraska Press, 1985. This is the essential history of the North American forests.

Dawson, William R. *et al. Report of the Advisory Panel on the Spotted Owl*. New York: National Audubon Society, 1986. Audubon Conservation Report No. 7. Applying the lessons from other endangered species, a panel of experts offers an excellent overview of the owl's precarious status and a plan for its survival.

Forest Watch. "The citizens' forestry magazine" is a unique tool for activists and citizens who want to understand the national forest planning process. For information, contact Cascade Holistic Economic Consultants, P.O. Box 3479, Eugene, Oregon 97403, (503) 686-CHEC.

Franklin, Jerry F. *et al. Ecological Characteristics of Old-Growth Douglas-Fir Forests*. U.S. Department of Agriculture Forest Service, Pacific Northwest Forest and Range Experiment Station, February 1981. General Technical Report PNW-118. Still the definitive scientific description of the old-growth ecosystem, this is easily understood by the general reader.

Harris, Larry D. *The Fragmented Forest: Island Biogeography Theory and the Preservation of Biotic Diversity*. Chicago: The University of Chicago Press, 1984. Explores the perils of fragmenting forest habitat and proposes solutions.

Kelly, David and Braasch, Gary. *Secrets of the Old Growth Forest*. Salt Lake City: Gibbs Smith, Publisher, 1988. This passionate defense of the ancient forests features the most spectacular assemblage of forest photos to appear in print.

Ketchum, Robert Glenn and Ketchum, Carey D., with photographs by Robert Glenn Ketchum. *The Tongass: Alaska's Vanishing Rain Forest*. New York: Farrar, Straus and Giroux, 1987. This lushly illustrated volume documents the assault on the old-growth forests of southeastern Alaska.

Maser, Chris. *Forest Primeval: The Natural History of an Ancient Forest*. San Francisco: Sierra Club Books, 1989. The forest is the protagonist of this "novel" spanning 1,000 years of natural history and human incursion.

------. *The Redesigned Forest*. San Pedro: R. & E. Miles, 1988. Maser, a mammalian biologist, describes the old-growth ecosystem and explains why present-day forestry may be a dead-end endeavor.

247

Norse, Elliott A. *Ancient Forests of the Pacific Northwest: Sustaining Biological Diversity and Timber Production in a Changing World.* Covelo, California: Island Press, 1989. The Wilderness Society's chief ecologist explores the old-growth ecosystem in an era of global warming.

O'Toole, Randal. *Reforming the Forest Service.* Washington, D.C., and Covelo, California: Island Press, 1988. O'Toole argues that the Forest Service puts timber sales first in order to boost its budget, and he offers a radical solution.

Pyle, Robert Michael. *Wintergreen: Listening to the Land's Heart.* Boston: Houghton Mifflin Company, 1988. Originally under the subtitle *Rambles in a Ravaged Land,* this is an ecologist's highly personal description of the natural world following the destruction of southwest Washington's old-growth forests.

Raphael, Ray. *Tree Talk: The People and Politics of Timber.* Covelo, California: Island Press, 1981, 1987. Essentially a series of interviews, this book provides an excellent and balanced overview of forest management.

Stewart, Hilary. *Cedar: Tree of Life to the Northwest Coast Indians.* Seattle and London: University of Washington Press, 1984. Through words and her own illustrations, Stewart shows how the traditional cultures of the Northwest made use of the region's most special tree.

Whitney, Stephen. *Western Forests.* The Audubon Society Nature Guides. New York: Alfred A. Knopf, Inc., 1986. The special bonus of this richly illustrated field guide is its detailed discussion of the forest ecosystems of the West.

PHOTO AND ART CREDITS

Unless otherwise noted, all photographs are by the author.
The maps on page 19 and the drawings on page 76 are by Linda Wilkinson.
Cover photo: Bob and Ira Spring
Page 38 top: Jimi Lott (photo first appeared in *The Seattle Times*)
Page 38 bottom: Adrian Dorst
Page 56 top: John D. Cress, courtesy of Weyerhaeuser Company archives
Page 56 bottom: courtesy of Weyerhaeuser Company archives
Page 93: Jeff Hughes, Alaska Department of Fish and Game
Page 130: courtesy of Weyerhaeser Company archives
Page 145: courtesy of Weyerhaeuser Company archives
Page 164: Daryl Williams, courtesy of the Tulalip Tribes
Page 207: Tracy L. Fleming

Notes

Abbreviations

CHEC = Cascade Holistic Economic Consultants
DEIS = Draft Environmental Impact Statement
DSEIS = *Draft Supplement to the Environmental Impact Statement for an Amendment to the Pacific Northwest Regional Guide*
FSEIS = *Final Supplement to the Environmental Impact Statement for an Amendment to the Pacific Northwest Regional Guide*
GTR = General Technical Report
MBS Plan = *Proposed Land and Resource Management Plan, Mount Baker-Snoqualmie National Forest*
NAS = National Audubon Society
NF = National Forest
NCASI = National Council of the Paper Industry for Air and Stream Improvement, Inc.
NIFM = Northwest Independent Forest Manufacturers
OSU = Oregon State University
PNWF&RExSta = Pacific Northwest Forest and Range Experiment Station
PNWReg = Pacific Northwest Region
PNWResSta = Pacific Northwest Research Station
TWS = The Wilderness Society
USFS = U.S. Department of Agriculture Forest Service
USFWS = U.S. Interior Department Fish and Wildlife Service
UW = University of Washington
WDNR = Washington Department of Natural Resources
WDT&ED = Washington Department of Trade and Economic Development
WDW = Washington Department of Wildlife

Unless otherwise noted, all interviews with the author were conducted between June 1985 and March 1989.

CHAPTER 1

Observations about forest ecology are drawn in large part from interviews with Franklin, from Franklin's address to the Ancient Forests Conference in Portland, Oregon, September 23, 1988, and from Jerry F. Franklin *et al.*, *Ecological Characteristics of Old-Growth Douglas-Fir Forests* (USFS, PNWF&RExSta, GTR PNW-118, February 1981).

Page 10. *"They're putting people . . . sheer stupidity."* *Seattle Post-Intelligencer*, May 1, 1985, p. A8.

Page 11. *Less than one-fifth . . . still stands.* Jack Ward Thomas *et al.*, "Management and Conservation of Old-Growth Forests in the United States," *Wildlife Society Bulletin*, 1988, Vol. 16, No. 13, pp. 252-62.

Page 11. *. . . 48,000 acres of virgin forests are being cleared.* Also *Over a million acres . . . is being preserved . .* USFS, PNWReg, *FSEIS*, 1988. Vol. 1, pp. III-31, III-41, IV-5. Acres are based on suitable spotted owl habitat.

Page 11. *The U.S. Bureau of Land Management . . . 22,000 acres.* Marvin L. Plenert, *Revised Finding of Northern Spotted Owl Listing Petitions.* U.S. Fish and Wildlife Service, April 25, 1989, p. 14.

Page 11. *. . . 850 million acres . . . virgin forests.* Marion Clawson, "Forests in the Long Sweep of American History," *Science*, June 15, 1979, p. 1169.

Page 11. *. . . "the world center of mushroom diversity."* Robert Michael Pyle, *Wintergreen: Listening to the Heart's Land.* Boston: Houghton Mifflin Company, 1988, p. 65. Originally published by Charles Scribner's Sons, 1986.

Page 14. *. . . one-fourth of total precipitation.* R. Dennis Harr, "Fog Drip in the Bull Run Municipal Watershed Oregon," *Water Resources Bulletin*, Vol. 18, No. 5, October 1982, pp. 785-8.

Page 14. *"These trees . . . whales of the forests."* Elliott Norse, address to Ancient Forests Conference, Portland, Oregon, September 24, 1988.

Page 14. *Only the huge eucalyptus . . . any ecosystem on earth.* J.F. Franklin and R.H. Waring, "Distinctive Features of the Northwestern Coniferous Forest," in *Forests: Fresh Perspectives from Ecosystem Analysis.* Corvallis, Ore.: OSU Press, 1979, pp. 60-61.

Page 14. *Conifers . . . a thousand to one.* Also *Seattle typically . . . Medford, Oregon.* Jerry F. Franklin and C.T. Dyrness, *Natural Vegetation of Oregon and Washington.* Corvallis: OSU Press, 1988, p. 53.

Page 16. *By 1989, Franklin . . . forest's ecological structure.* Franklin, address to symposium, Old-Growth Douglas-fir Forests: Wildlife Communities and Habitat Relationships, Portland, March 29, 1989.

Page 17. *"For so large a tree . . . covered the land."* John Muir, *Steep Trails*, William Frederic Bade, ed. Boston: Houghton Mifflin Co., 1918, p. 2.

Page 17. *the alaska yellow cedar . . . 3,500 years.* R.H. Waring and J.F. Franklin, "Evergreen Forests of the Pacific Northwest," *Science*, June 29, 1979, p. 1382.

Page 18. *Sitka spruce . . . less than thirty-five years.* Stephen F. Arno and Ramona P. Hammerly, *Northwest Trees.* Seattle: The Mountaineers, 1977, p. 56.

Page 20. *More than 130 epiphytic mosses . . "in their own crowns."* Nalini M. Nadkarni, "Roots That Go Out on a Limb," *Natural History*, February 1985, pp. 43-8.

Page 21. *"Most life . . . reached the forest floor."* Marston Bates, *The Forest and the Sea.* New York: Signet Science Library Books, 1964, p. 21. Originally published by Random House, Inc., 1960.

Page 21. *"Can old growth . . . old-growth stands."* Society of American Foresters, *Scheduling the Harvest of Old Growth.* Bethesda, Md., 1984, p. 31.

Pages 22-23. *. . . the class of fungi known as mycorrhizae.* Two excellent discussions of the truffle-mammal-tree relationship can be found in Trappe and Chris Maser, "Ectomycorrhizal Fungi: Interactions of Mushrooms and Truffles with Beasts and Trees," in *Mushrooms and Man: An Interdisciplinary Approach to Mycology*, Tony Walters, ed., (USFS, 1977), pp. 165-76, and in Maser, Trappe, and Douglas C. Ure, "Implications of Small Mammal Mycophagy to the Management of Western Coniferous Forests," in *Transactions of the 43rd North American Wildlife and Natural Resources Conference* (Washington, D.C.: Wildlife Management Institute, 1978), pp. 78-88.

Page 23. *"Thus . . the fungus thrives."* Chris Maser and James M. Trappe, tech. eds., *The Seen and Unseen World of the Fallen Tree*. USFS, PNWF&RExSta, GTR PNW-164, March 1984, p. 31.

Page 23. *Maser . . . parts of Europe.* Chris Maser, *The Redesigned Forest*. San Pedro, Calif.: R. & E. Miles, 1988, pp. 69-82.

Page 23. *Other scientists . . . "genetically improved" trees.* Roy R. Silen, *Nitrogen, Corn, and Forest Genetics: The Agricultural Yield Strategy -- Implications for Douglas-fir Management*. USFS, PNWF&RExSta, GTR PNW-137, June 1982.

Page 23. *Global deforestation . . . next several decades.* Peter H. Raven, address to NAS biennial convention, Bellingham, Washington, August 24, 1987.

Page 23. *. . . bark of the Pacific yew . . . tumors in mice.* *The New York Times*, May 3, 1987, p. 29.

Page 24. *More than one-third . . . wood products.* Con Schallau, Douglas Olson and Wilbur Maki, "An Investigation of Long-Term Impacts on the Economy of Oregon of Alternative Timber Supply Forecasts," in Gary J. Lettman, tech. ed., *Assessment of Oregon's Forests*. Salem: Oregon Department of Forestry, July 1988, p. 184.

Page 25. *. . . two-fifths . . . is sold abroad.* Debra D. Warren, *Production, Prices, Employment, and Trade in Northwest Forest Industries, Third Quarter 1988*. USFS, PNWResSta, Resource Bulletin PNW-RB-162, January 1989, pp. 13, 29-33.

Page 26. *. . . Robert Vincent . . . "a thing of beauty."* Comments to a conference sponsored by League of Women Voters of Central Lane County and Bureau of Governmental Research and Service, University of Oregon. In *Old Growth Forests: A Balanced Perspective*. Eugene: Bureau of Governmental Research and Service, 1982, pp. 39-40.

CHAPTER 2

The description of *seyowen* is drawn in large part from Astrida R. Blukis Onat and Jan L. Hollenbeck, editors, *Inventory of Native American Religious Use, Practices, Localities and Resources*. Seattle: Mount Baker-Snoqualmie NF and Institute of Cooperative Research, Inc., April 1981. Also see Pamela Amoss, *Coast Salish Spirit Dancing: The Survival of an Ancestral Religion*. Seattle: UW Press, 1978.

Pages 29-30. *It provided clothing . . . hollowed out and expanded.* Hilary Stewart, *Cedar: Tree of Life to the Northwest Coast Indians*. Seattle: UW Press, 1984.

Pages 31. *Scientists do not know . . . like those of today.* *The Seattle Times*, October 5, 1986, p. J6. Wayne Suttles, in *Coast Salish Essays* (Seattle: UW Press, 1987), pp. 265-81), argues for a coastal migration route.

Page 31. *In a provocative . . . "totem poles became available."* Richard J. Hebda and Rolf W. Mathewes, "Holocene History of Cedar and Native Indian Cultures of the North American Pacific Coast," *Science*, Vol. 225, August 17, 1984, pp. 711-3. The evolution of the coastal forests in response to climate changes is explained in the *Encyclopedia of American Forest and Conservation History*, Vol. 1, pp. 150-7.

Page 31. *Ethnographer Wayne Suttles . . . spruce in the north.* Suttles, *Coast Salish Essays*, pp. 36-8.

Page 32. *. . . population . . . in the millions.* Astrida R. Blukis Onat, "Cultural Control and Management of Resources in the North American Northwest Coast and in the North European Baltic Areas: A Comparison." Unpublished manuscript.

Page 32. *The Hesquiat Indians . . . turned into cedars.* Stewart, *Cedar*, p. 27.

Page 32. *. . . "by nature . . . besetting sin."* Hu Maxwell, quoted in Calvin Martin, "Fire and Forest Structure in the Aboriginal Eastern Forest," *The Indian Historian*, Vol. 6, No. 4, Fall 1973, p. 40.

Page 32. *. . . Whidbey Island . . . "proved so successful."* Richard White, *Land Use, Environment, and Social Change: The Shaping of Island County, Washington*. Seattle: UW Press, 1980, pp. 21-25, 159.

Page 36. *". . . hippies . . . with binoculars."* Sam Cagey, interview with author.

Pages 39-41. *Do the tribes . . . "no teeth in it."* The legal background here is drawn from three court decisions: U.S. Supreme Court Case No. 86-1013, *Richard E. Lyng, Secretary of Agriculture, et al., Petitioners v. Northwest Indian Cemetery Protective Association et al.*, decision of April 19, 1988; U.S. Court of Appeals, Ninth Circuit, in *Northwest Indian Cemetery Association et al. v. R. Max Peterson et al.*, Case Nos. C-82-4049 SAW and C-82-5943 SAW, decision of May 24, 1983; and Ninth Circuit Case Civ A. No. 83-2225, decision of July 22, 1986.

Pages 41-42. *We shouldn't be . . . "it was cogent indeed."* Calvin Martin, "The American Indian as Miscast Ecologist," *The History Teacher*, Vol. XIV, No. 2, February 1981, p. 246.

Page 42. *The Indian Tribes of British Columbia . . . South Moresby National Park.* *The Seattle Times*, December 2, 1985, p. B1; June 23, 1987, p. B1; and July 8, 1987, p. D2.

Page 42. *. . . victory was overshadowed . . . Stein Valley.* "Canada's vanishing forests," *Maclean's*, January 14, 1985, pp. 36-42; "A Threat to Tall Timber," *Sierra*, November/December 1988, pp. 100-103; and "Stein Valley: The choice is ours," Western Canada Wilderness Committee, October 1987.

Page 42. *Because of the Salish . . . agreed to sponsor the study.* Kurt Russo, interview with author.

Pages 42-43. *The results of the study . . . shift from one spot to another.* Blukis Onat and Hollenbeck, *Inventory of Native American Religious Use*.

Pages 43-44. *When the Mount Baker-Snoqualmie . . . essential to their religion.* USFS, PNWReg, *MBS Plan*, December 1987, p. 3-4, and *DEIS, MBS Plan*, p. IV-73.

Page 44. *In a 1985 article . . . "They have lost it."* *The Seattle Times*, August 23, 1985, p. A1.

Page 45. *"From close observation . . . among the younger Indians."* Blukis Onat and Hollenbeck, *Inventory of Native American Religious Use*, p. 144.

Page 45. *In 1921, . . . "Indian offences."* Office of Indian Affairs Circular No. 1665, April 26, 1921. Quoted in Jon Magnuson, "Affirming Native Spirituality: A Call to Justice," *The Christian Century*, December 9, 1987, p. 1116.

Page 46. *. . . upwards of 5,000 adherents . . .* Pamela Amoss, testimony in U.S. District Court, Seattle, *U.S. v. Cooper et al.*, March 1983.

Page 46. *. . . eight Protestant denominations . . . sacred sites on public lands.* The Rev. Thomas L. Blevins *et al.*, "A Public Declaration to the Tribal Councils and Traditional Spiritual Leaders of the Indian and Eskimo Peoples of the Pacific Northwest," Thanksgiving Day, 1987.

CHAPTER 3

The discussion of Americans and their forests in the colonial period is drawn primarily from two excellent sources: Richard G. Lillard, *The Great Forest* (New York: Alfred A. Knopf, 1947), and Thomas R. Cox *et al.*, *This Well-Wooded Land: Americans and Their Forests from Colonial Times to the Present* (Lincoln: University of Nebraska Press, 1985).

Page 49. *Christopher Columbus . . . "with leaves never shed."* Rutherford Platt, *The Great American Forest*. Englewood Cliffs, N.J.: Prentice-Hall, Inc., 1969, p. 11.

Page 51. *. . . John Ervin . . . "mud Chimneys."* Sam J. Ervin, Jr., "Entries in Colonel John Ervin's Bible," *South Carolina Historical Magazine*, July 1978, pp. 219-228.

Pages 51-52. *. . . Puritans of New England . . . secure in the forest.* Roderick Nash, *Wilderness and the American Mind*. New Haven and London: Yale University Press, revised edition, 1973, pp. 1-7, 23-43.

Page 52. *For every acre . . . make way for farms.* Cox *et al.*, p. 103.

Page 52. *"We have the directive. . . changes occur."* H.D. Bennett, quoted in Jack Shepherd, *The Forest Killers*. New York: Weybright and Talley, 1975.

Page 53. *This held true . . . two and a half centuries.* Michael Williams, "Clearing the United States forests: pivotal years 1810-1860," *Journal of Historical Geography*, Vol. 8, No. 1, 1982, pp. 12-28.

Page 53. *Thomas Jefferson's . . . "waste as we please."* Shepherd, *The Forest Killers*.

Pages 54-55. *Chief Seattle's . . . "hidden from us."* Seattle's words were doubtless embellished by his interpreter, notes David Buerge in "The man we call Seattle," *The Weekly*, Seattle, June 29-July 5, 1983, pp. 24-27.

Page 55. *. . . George Vancouver . . . "spars the world produces."* Vancouver, *The Voyage of George Vancouver, 1791-1795, Volume II. A Voyage of Discovery to the North Pacific Ocean and Round the World*, W. Kaye Lamb, ed. London: The Hakluyt Society, 1984.

Page 55. *Meriwether Lewis . . . "any other species."* Lewis, *The Expeditions of Lewis and Clark, Vol. II*. Reprint of 1814 report. Ann Arbor, Mich.: University Microfilms, Inc., 1966, p. 55.

Page 55. *Botanist David Douglas . . . "objects in Nature."* Athelstan George Harvey, *Douglas of the Fir: A Biography of David Douglas, Botanist*. Cambridge: Harvard University Press, 1947, p. 57.

Page 57. *Stewart H. Holbrook . . . "daylight into* this *swamp."* Holbrook, *The American Lumberjack*. New York: Collier Books, enlarged edition, 1972, p. 147. Originally published as *Holy Old Mackinaw* by The Macmillan Company, 1938.

Pages 57-59. *Known to his employees . . . to the Pacific Northwest.* This account of Weyerhaeuser and his move to the Pacific Northwest is drawn principally from Lillard, pp. 197-208.

Page 59. *One thirty-six-square-mile township . . . any other port before or since.* Edwin Van Syckle, *They Tried To Cut It All*. Seattle: Pacific Search Press, 1980, pp. 127-9.

Pages 59-60. *As the logging business . . . pull logs to the landings.* For more on the rough-and-tumble logging days, see Holbrook and Van Syckle.

Pages 60-61. *Haywire Tom Newton . . . Grand Coulee Dam.* Ruby El Hult, *The Untamed Olympics: The Story of a Peninsula*. Portland, Oregon: Binfords & Mort, 1971, pp. 194-6.

Page 61. *. . . nearly a billion acres of natural forest . . .* Marion Clawson, "Forests in the Long Sweep of American History," *Science*, Vol. 204, June 15, 1979, p. 1168.

Page 61. . . . *"Forestry begins with the ax."* William B. Greeley, *Forests and Men*. Garden City, N.Y.: Doubleday & Co., 1951, p. 229.

Pages 61-62. *Early in this century . . . "standards of commerce."* Greeley, pp. 116-7.

Page 62. *The U.S. merchant marine . . . foreign-flag vessels.* Cox *et al.*, p. 114.

Page 62. *The Swamp Land Act . . . land reclamation.* Lillard, p. 177.

Pages 64-65. *The two men . . . "right there as we did."* Dyan Zaslowsky and The Wilderness Society, *These American Lands*. New York: Henry Holt and Company, 1986, pp. 78-9.

Page 65. . . . *"[T]he noble giant . . . falls to earth."* Muir, *Steep Trails*, p. 224.

Page 65. *"With the extirpation . . . diminution of them."* George Perkins Marsh, *The Earth as Modified by Human Action*. A last revision of *Man and Nature*. New York: Charles Scribner's Sons, 1898, pp. 292-3, 356.

Page 66. *A few once-rich forests . . . once stood there. Los Angeles Times*, June 14 and 21, 1987, p. 1.

Page 66. . . . *fifty acres of tropical jungle disappear every minute. The Seattle Times*, August 19, 1988, p. A3.

Page 66. *The old-growth forests . . . 190 acres a day . . .* Computed from figures in USFS, PNWReg, *FSEIS*, Vol. 1, p. IV-5; and Plenert, *Revised Finding*.

Pages 66-67. *Dr. Seuss . . . "may come back.* Dr. Seuss (Theodor Seuss Geisel), *The Lorax*. New York: Random House, Inc., 1971.

CHAPTER 4

Background information on the marbled murrelet is drawn primarily from David B. Marshall, *Status of the Marbled Murrelet in North America, with Special Emphasis on Populations in Washington, Oregon and California* (Unpublished report to Audubon Society of Portland, January 4, 1988).

The use of old-growth forests by Alaska black-tailed deer is documented in papers by Cathy L. Rose and by Susan K. Stevenson and James A. Rochelle in *Fish and Wildlife Relationships*, William R. Meehan *et al.*, eds. (American Institute of Fishery Research Biologists, December 1984, pp. 285-90, 391-6). The ancient murrelet is discussed by Donald A. Blood and Gary G. Anweiler in the same volume, pp. 297-302.

Information about the ecological role of snags and logs is drawn primarily from Franklin *et al.*, *Ecological Characteristics*, from E. Reade Brown, tech. ed., *Management of Wildlife and Fish Habitats in Forests of Western Oregon and Washington* (USFS, PNWReg, Publication No. R6-F&WL-192-1985, Vol. I, pp. 129-170), and from Maser and Trappe, eds., *The Seen and Unseen World of the Fallen Tree*.

The unique food chain of old-growth forests is described by Larry D. Harris, Chris Maser, and Arthur McKee in "Patterns of Old Growth Harvest and Implications for Cascades Wildlife," in *Transactions of the 47th North American Wildlife and Natural Resources Conference*, 1982, and Maser and Harris in Harris, *The Fragmented Forest* (Chicago: University of Chicago Press, 1984).

Background information on the pileated woodpecker and pine marten, as well as criticism of the Forest Service's use of "habitat areas," can be found in NCASI, *Review of the Minimum Management Requirements for Indicator Species: Pine Marten and Pileated Woodpecker* (NCASI Tech. Bull. No. 522. New York: April 1987). Larry Irwin's suggestion that "suboptimal habitat" be provided in managed forests can be found in this publication and in *Ecology of the Spotted Owl in Oregon and Washing-*

ton (NCASI Tech. Bull. No. 509, November 1986). Irwin also discussed this idea before Washington Wildlife Commission, January 15, 1988, and in an interview with the author.

Page 72. *Rutherford Platt . . . "stalking and cover."* Platt, *The Great American Forest*. Englewood Cliffs, N.J.: Prentice-Hall, Inc., 1969, p. 172.

Page 72. *. . . complaint of a pioneer . . . "recall a songbird."* Murray Morgan, *The Last Wilderness*. Seattle: UW Press, 1980, p. 7.

Page 73. *John Muir . . . "larger than mosquitoes."* Muir, *Steep Trails*, p. 325.

Page 74. *Explorer Meriwether Lewis . . . "in part sound."* Lewis, *The Expeditions of Lewis and Clark, Vol. II*, p. 79.

Page 77. *In a study . . . "small percentage of the time."* Peter H. Morrison and Frederick J. Swanson, *Fire History and Patterns in Cascade Mountain Landscape*. USFS, PNWF&RExSta, GTR, in press.

Page 79. *Thirty thousand deerskins . . . years between 1880 and 1910."* John W. Schoen *et al.*, "Wildlife-Forest Relationships: Is a Reevaluation of Old Growth Necessary?" in *Transactions of the 46th North American Wildlife Conference*, 1981, pp. 533-4.

Page 79. *Up to 50,000 deer . . . severe winter.* TWS, *America's Vanishing Rain Forest: A Report on Federal Timber Management in Southeast Alaska*. April 1986, p. 158.

Pages 80-81. *By 1974, Wight . . . "changes in the habitat."* Howard M. Wight, "Nongame Wildlife and Forest Management," in *Wildlife and Forest Management in the Pacific Northwest*, Hugh C. Black, ed. Corvallis: Forest Service Laboratory and School of Forestry, OSU, December 1974, pp. 27-38.

Page 87. *. . . ivory-billed woodpecker . . . found in Cuba.* Robert R. Dawson *et al.*, *Report of the Advisory Panel on the Spotted Owl*. New York: NAS, Audubon Conservation Report No. 7, 1986, p. 22. Also *Newsweek*, May 19, 1986, p. 53.

Page 87. *. . . the fisher . . . coastal Oregon and Washington.* Harris, *The Fragmented Forest*, p. 83. Also Keith Aubry, interview with author.

Page 87. *The bald eagle . . . old-growth fir forest.* Brown, ed., *Management of Wildlife and Fish Habitats*, p. 273. Also *The Seattle Times*, January 13, 1987, p. B2.

Page 88. *Environmentalists . . . its nest.* Bill Carroll, "Are the Texas Forests Planning Extinction for the Red-Cockaded Woodpecker?" *Forest Watch*, April 1988, pp. 23-7.

Pages 88-89. *In the Northwest . . . logging plans.* USFS, *Final Supplement to the EIS*, Vol. 1, pp. IV-35 to IV-41.

Page 89. *In the 1983 windstorm . . . roads or clearcuts.* Jerry F. Franklin and Richard T.T. Forman, "Creating Landscape Patterns by Forest Cutting: Ecological consequences and principles," *Landscape Ecology*, Vol. 1, No. 1, 1987, p. 11.

Page 90. *Scouring the records . . . according to Newmark.* William D. Newmark, "A Land-Bridge Island Perspective on Mammalian Extinctions in Western North American Parks," *Nature*, January 29, 1987, pp. 430-2.

Page 90. *James F. Quinn . . . present in the parks.* Quinn *et al.*, "Mammalian Extinctions from National Parks in the Western United States," *Ecology*, in press.

Page 90. *In a separate article . . . periphery of the parks.* James F. Quinn, "Extinction Rates and Species Richness of Mammals in Western North American Parks," *Biological Conservation*, in press.

Page 91. *"The answer . . . produce more."* Jack Ward Thomas *et al.*, "Management and Conservation of Old-Growth Forests in the United States," *Wildlife Society Bulletin*, 1988, Vol. 16, No. 13, pp. 252-62.

Pages 91-93. *Bill Hagenstein . . . "no people around."* Nancy Wood, *Clearcut: The Deforestation of America.* San Francisco: Sierra Club, 1971.

Page 93. *Gus Kuehne . . . "useful products."* Seattle Post-Intelligencer, October 13, 1988, p. A13.

Page 94. *In his essay . . . "such a view."* Aldo Leopold, *A Sand County Almanac, With Essays on Conservation from Round River.* New York: Bal-lantine Books, 1977. Originally published by Oxford University Press, 1966.

CHAPTER 5

Pages 97-99. *"I believe that the reserve . . . no other reason."* Elizabeth S. Poehlman, *Darrington: Mining Town/Timber Town.* Kent, Washington: Gold Hill Press, 1979, p. 137.

Page 100. *As many as 100 . . . "SPOTTED OWL STEW 99¢."* The Seattle Times and Seattle Post-Intelligencer, February 27, 1988.

Page 100. *Seven months after . . . "our picture in the post office."* Jay Simons, "The Great Northwest Log Haul," *American Forests,* September/October 1988, Vol. 94, Nos. 9 & 10, pp. 17-20.

Page 103. *. . . 18 million acres of managed second growth forest . . .* Second-growth acreage was computed by adding Oregon state figures for nonfederal timber-land, Washington state figures for nonfederal commercial forest land, national forest land "technically suitable" for timber production but not suitable as spotted owl habitat, and an estimated one million acres of second growth managed by BLM. Lettman, "Timber Summary," in *Assessment of Oregon's Forests,* p. 90; Larsen and Wadsworth, *Washington Forest Productivity Study,* p. 1; and USFS, PNWReg, *FSEIS,* Vol. I. p. III-32.

Page 103. *. . . 1.1 million acres in wilderness and national parks.* Figure represents "currently suitable" spotted owl habitat. USFS, PNWReg, *FSEIS,* Vol. I, pp. III-31 and III-41.

Page 103. *. . . 4 million acres of virgin forests . . .* The Forest Service's *FSEIS* (pp. III-31, III-32, III-41) show 5.5 million acres of spotted owl habitat in parks, wilderness, and commercial timberland owned by the federal government. those figures don't reflect recent logging, however, and only a portion of the remainder could be considered old growth.

Page 104. *. . . stumpage prices . . . in just three years.* Richard W. Haynes, *Inventory and Value of Old-Growth in the Douglas-Fir Region.* USFS, PNWResSta, Research Note PNW-437, January 1986, p. 12. Prices are for 1980 and 1983.

Page 105. *A hundred lumber . . . during the early eighties.* George H. Weyerhaeuser, address to Portland Rotary Club, June 9, 1987.

Page 105. *. . . and another 45 . . . 1987 and 1989.* Rep. Jolene Unsoeld, quoted in *Seattle Post-Intelligencer,* April 14, 1989, p. C1.

Page 105. *Workers had helped . . . wages to previous levels.* Jones interview, June 13, 1988. Also *The Darrington Dispatch,* August 1988.

Page 105. *As the strike dragged on . . . with nonunion labor.* Seattle Post-Intelligencer, December 24, 1988, p. A1.

Page 106. *To finance its buyout . . . destruction of the forests.* Also, *"They're just leveling everything" . . . Pacific Lumber employee.* The New York Times, March 2, 1988, p. A16. Also see *Newsweek,* July 6, 1987, p. 38.

Pages 106-107. *. . . Bruce Engel's . . . modest debt burden.* The Seattle Times, p. C1, and Seattle Post-Intelligencer, June 15, 1987, p. B9.

Page 107. *Oregon and Washington lumber . . . eight-year period.* Warren, *Production, Prices, Employment, and Trade,* p. 17.

Page 107. *Weyerhaeuser coupled . . . a senior Weyerhaeuser executive.* Robert L.

Schuyler, address to paper and forest products analysts, New York, September 16, 1987.

Page 107. *The number of timber-related jobs . . . 64,000 by the year 2000.* Brian R. Wall, *Employment Implications of Projected Timber Output in the Douglas-fir Region, 1970-2000* (USFS, PNWF&RExSta, November 1973). Also Mark Houser, unpublished paper, November 1984.

Page 107. *. . . a loss of just under 5,000 timber-related jobs. The Seattle Times*, March 3, 1988. The analysis was presented by David D. Green and C.W. Corssmit to the Economic Impact Analysis Conference, Portland, March 2, 1988.

Page 107. *A 1988 study . . . would be lost anyway.* Jeffrey T. Olson, *Pacific Northwest Lumber and Wood Products: An Industry in Transition*. Washington, D.C.: TWS and National Wildlife Federation, *National Forests: Policies for the Future, Vol. 4*, p. vii.

Page 109. *Log shipments . . . except 1979.* Warren, *Production, Prices, Employment, and Trade*, p. 20.

Page 109. *Exports hit a new high in 1988.* Log exports were 3.7 billion board feet, or 25 percent of Oregon and Washington's harvest. *Seattle Post-Intelligencer*, March 20, 1989, p. B5.

Page 110. *Weyerhaeuser . . . exports go to Japan.* Tom Ambrose, interview with author, February 5, 1988.

Page 110. *The company's shipments . . . national forests combined.* Weyerhaeuser and ITT Rayonier exported 544 million board feet from Grays Harbor in 1987, while 538 million board feet were cut on the Olympic and Mount Baker-Snoqualmie Forests. Interviews with Karl Wallin and Ron DeHart, and Olympic NF, *1987 Annual Report*, p. 4.

Page 110. *. . . a whopping $100 more . . . one public auction . . . Seattle Post-Intelligencer*, November 12, 1987, p. A10.

Page 110. *The secretary of agriculture . . . substitutions take place.* U.S. General Accounting Office, *Potential Impacts of Tighter Forest Service Log Export Restrictions*. GAO/RCED-85-17, January 28, 1985.

Page 111. *Washington . . . overseas in 1987.* Warren, *Production, Prices, Employment, and Trade*, pp. 13, 20, 32.

Page 111. *At least two-thirds . . . through local mills.* Gus Kuehne, interview with author, January 6, 1988.

Page 111. *"Their footprints were all over this." The Seattle Times*, July 14, 1988, p. A13.

Page 112. *One securities analyst . . . 10-percent profit making lumber.* Houser, November 1984.

Page 113. *A 1981 study . . . jobs in Washington.* Northwest Economic Consultants, Wesley Rickard, Inc., Richard W. Parks and Judith Cox, and Ed Williston Associates, Inc., *Revenue and Job Impacts of a Ban on Log Exports from State-Owned Lands in Washington*. Prepared for Washington Citizens for World Trade, August 1981, p. iii.

Page 113. *Northwest Independent . . . created in Washington.* Also Page 36, *NIFM pointed . . . lumber exporter.* NIFM, "The Wise Use of Our State Timber." Unpublished paper, June 22, 1988, pp. 4 and 7.

Page 113. *At Vanport . . . Japanese builders. The Seattle Times*, March 22, 1987, p. D1.

Page 113. *At the new Pacific Veneer . . . specifications. The Seattle Times*, December 16, 1987, p. H1.

Pages 113-114. *The Evergreen Partnership . . . "1.2 million units."* Greg Shellberg, interview with author, February 5, 1988; and *Puget Sound Business Journal*, November 21, 1988, p. 1.

Page 117. *On a foggy . . . than he intended.* Morris "Babe" Giebel, interview with author, October 23, 1985.

CHAPTER 6

Page 119. *Edward P. Cliff . . . "prelude" to progress.* Michael Frome, *The Forest Service*. Boulder, Colo.: Westview Press, 1984, p. 109.

Page 119. *William B. Greeley . . . "Dante's* Inferno." Greeley, *Forests and Men*, p. 130.

Pages 119-120. *But as the war ended . . . sustained-yield unit.* David A. Clary, *Timber and the Forest Service*. Lawrence: University Press of Kansas, 1986, pp. 127-31.

Page 120. *. . . dropped from 96 million . . . one million in 1987.* Rick Hall, interview with author, April 28, 1989.

Page 120. *"Forestry is Tree Farming . . . regulated grazing."* Gifford Pinchot, *Breaking New Ground*. Seattle: UW Press Americana Library edition, 1972, p. 31.

Page 121. *President Theodore Roosevelt . . . "Asia Minor is today."* Robert H. Nelson, "Methodology Instead of Analysis: The Story of Public Forest Management," in Robert T. Deacon and M. Bruce Johnson, eds., *Forestlands: Public and Private*. San Francisco: Pacific Institute for Public Policy Research (Harper & Row), 1985.

Page 121. *"There is no reason . . . hold out indefinitely."* Clary, *Timber and the Forest Service*, p. 18.

Page 122. *"The experience . . . facts of supply and demand."* Sherry H. Olson, *The Depletion Myth: A History of Railroad Use of Timber*. Cambridge: Harvard University Press, 1971, p. 178.

Pages 122-123. *"The export policy . . . tear this community apart."* Seattle Post-Intelligencer, March 7, 1989, p. A10.

Page 123. *Nearly two-thirds . . . 19 years old.* Donald R. Gedney, "The Private Timber Resource," in *Assessment of Oregon's Forests*, p. 55.

Page 123. *. . . twenty-two million acres . . . western Oregon.* Western Washington 10.2 million acres, western Oregon, 11.6 million acres. David N. Larsen, "Washington's Timber Situation," WDNR, unpublished paper, August 14, 1987. Also Gary J. Lettman, "Timber Summary," in *Assessment of Oregon's Forests*, p. 90.

Page 123. *A rather optimistic . . . increase by four percent.* David N. Larsen and Robert K. Wadsworth, *Washington Forest Productivity Study, Phase III, Part II*. WDNR, January 1982, p. 16.

Page 123. *When Oregon's . . . would never be recouped.* Lettman, "Timber Summary," in *Assessment of Oregon's Forests*, p. 89.

Page 123. *. . . 44-percent drop . . . Douglas fir region.* USFS, *An Analysis of the Timber Situation in the United States 1952-2030*. Forest Service Report No. 23, December 1982, p. 476.

Page 125. *Marion Clawson . . . "he might never see."* Clawson, "The National Forests," *Science*, Vol. 191, February 20, 1976, pp. 764-5.

Pages 125-126. *Federal prosecutors . . . lumber from the thief.* Seattle Post-Intelligencer, November 21, 1987, June 17, 1988, and October 18, 1988.

Page 126. *"Sure . . . He just got caught" . . .* Seattle Post-Intelligencer, June 14, 1988, p. A4.

Page 126. *The rain forests . . . future session of Congress.* A detailed history of the Alaska sweetheart deals and taxpayers' losses can be found in *America's Vanishing Rain Forest: A Report on Federal Timber Management in Southeast Alaska*. TWS, 1986.

Page 127. *When O'Toole examined . . . was sold below-cost.* Randal O'Toole and Dieter Mahlein, *Review of the Proposed Olympic Forest Plan and DEIS*. Eugene: CHEC, March 1987, pp. 4-7.

Pages 127-128. *O'Toole attributes . . . its budget.* O'Toole, *Reforming the Forest Service*. Covelo, Calif.: Island Press, 1988.

Page 128. *Robert H. Nelson . . . "a stabilizing influence."* Nelson, "Methodology Instead of Analysis," in *Forestlands: Public and Private*, Deacon and Johnson, eds.

Page 132. *In western Oregon . . . timberlands in 1987.* Gedney, "The Private Timber Resource," *Assessment of Oregon's Forests*, p. 55.

Page 132. *Virgin forests . . . as late as 1984.* Larsen, "Washington's Timber Situation," p. 4.

Pages 134-135. *The numbers show this . . . ten percent of the harvest.* Also Page 31-32. *While timber inventories . . . during the 1980s.* Larsen, "Washington's Timber Situation," figures 1 and 3 and Table 2.

Page 135. *More importantly, 44 percent . . . no better in Washington.* Philip L. Tedder, address to "Small Woodlands Contribution to Oregon" conference, Salem, May 1 and 2, 1987.

Pages 135-136. *In western Oregon . . . "timber at full potential."* John H. Beuter, K. Norman Johnson, and H. Lynn Scheurman, *Timber for Oregon's Tomorrow: An Analysis of Reasonably Possible Occurrences.* Corvallis: Forest Research Laboratory, School of Forestry, OSU, Research Bulletin No. 19, 1976, pp. 7, 23, 43.

Page 136. *In western Washington . . . Site Class 1 acres.* Larsen and Wadsworth, *Washington Forest Productivity Study*, pp. 42-43.

Page 136. *Marion Clawson . . . if not the present one.* Clawson, *The Economics of U.S. Nonindustrial Private Forests.* Washington, D.C.: Resources for the Future, 1979, p. 11.

Page 136. *The budget for cost-sharing . . . $9.7 million in 1986 . . .* Charlie Krebs, interview with author, February 11, 1988. Figures in clude funds administered by state and private forestry and by the Soil and Conservation Service.

Page 136. *Nixon's Advisory Panel . . . production on those lands.* Clark Shepard Binkley, *Timber Supply from Private Nonindustrial Forests.* New Haven: Yale University School of Forestry and Environmental Studies, Bulletin No. 92, 1981, p. 12.

Pages 136-137. *Washington's Department . . . timber base in western Washington.* Larsen and Wadsworth, *Washington Forest Productivity Study*, pp. 12-13.

Page 137. *Given the high quality . . . acts of 1984 have had.* The Washington Wilderness Act removed one million acres from the national forests' commercial timber base. Much of this land has no forest cover; the remainder is far less productive than the low-elevation private forests.

CHAPTER 7

Page 142. *Herbicides are applied annually to only 2.5 percent of private forests . . .* Gary J. Lettman, "Timber Management Practices in Oregon, in *Assessment of Oregon's Forests*, p. 63.

Page 142. *Neighbors . . . dropping the whole idea.* Seattle *Post-Intelligencer*, October 18, 1988, p. B2, and February 18, 1989, page B1.

Page 145. *Weyerhaeuser scientists . . . "over that of wild seed."* Farnum, Timmis, and Kulp, "Biotechnology of Forest Yield," *Science*, Vol. 219, February 11, 1983, p. 700.

Pages 148-149. *As recounted by Jack Doyle . . . "much grander scale in the future."* Jack Doyle, *Altered Harvest: Agriculture, Genetics, and the Fate of the World's Food Supply.* New York: Viking, 1985, pp. 12-17.

Page 150. *Addressing a crowd of environmentalists . . .* Chris Maser, address at

260   *Notes*

Portland State University, January 13, 1987.

Page 150. *It is worthy of note . . . "are seldom discussed."* Maser, *The Redesigned Forest*, p. 69.

Page 151. *"Our forestry . . . benefit of following generations."* Also, *One German scientist . . . from Czechoslovakia.* Maser, *The Redesigned Forest*, pp. 69, 82.

Pages 152-154. *Consider the arguments . . . "truly sustainable yields."* Roy R. Silen, *Nitrogen, Corn, and Forest Genetics.*

Pages 154-155. *Because the fungus . . . in affected areas.* E.M. Hansen *et al.*, "Biology and Management of Black-Stain Root Disease in Douglas-Fir," in *Forest Pest Management in Southwest Oregon*, O.T. Helgerson, ed. Corvallis: Forest Research Laboratory, OSU, 1986. Proceedings of a conference.

Page 155. Phytophthora lateralis . . . *hastening its demise.* Thomas Lawson, *Management of Port-Orford-Cedar and its Influence on Phytophthora Root Rot.* Eugene: 1983. CHEC Research Paper No. 13.

Page 155. *Spruce budworm . . . lower-value species.* O'Toole, *Reforming the Forest Service*, p. 87.

Page 155. *The red-cockaded woodpecker . . . predator on the pine beetle.* Edward C. Fritz, "Logging Fails to Control Pine Beetle," *Audubon Activist*, May 1987, p. 9. Page 33. Also O'Toole, p. 88.

Page 156. *"Current forest management . . . old growth is cut."* T.D. Schowalter and J.E. Means, "Pests Link Site Productivity to the Landscape," in D.A. Perry *et al.*, *Maintaining the Long-term Productivity of Pacific Northwest Forest Ecosystems*. Portland: Timber Press, in press.

Page 156. *A hot fire may volatilize 95 percent of the soil's nitrogen.* Roger Hart, "The Questionable Practice of Slash Burning, Part II," *NCAP News* (Northwest Coalition for Alternatives to Pesticides), Vol. 4 No. 4, Winter 1985.

Page 157. *. . . land-clearing fires . . . one-tenth of the carbon . . .* Seattle Post-Intelligencer, August 12, 1988, p. A6.

Page 158. *"It is actually fortunate . . . underneath partial shade."* O'Toole, p. 158.

Page 159. *"These trees . . . 'forest raper.'"* Shepherd, *The Forest Killers*, pp. 140-1.

Page 160. *Harris, bringing island . . . of mature forest.* Harris, *The Fragmented Forest*, pp. 108-65.

Page 160. *Maser, under . . . three 30s.* Maser, *The Redesigned Forest*, pp. 180-90.

Page 160. *Timber framers . . . for their market.* Timber Framers Guild of North America, "Also at Issue: Sustainable Old Growth," *American Forests*, September/October 1988, pp. 15-16, 78. Also Randal O'Toole, "Timber Framers Raise Old-Style Buildings and Old-Growth Consciousness," *Forest Watch*, November 1988, pp. 8-11.

Page 161. *"We, in western Oregon . . . through restoration forestry."* Maser, *The Redesigned Forest*, p. 186.

CHAPTER 8

Page 162. The First Salmon Ceremony at the Tulalip Reservation was held June 15, 1985.

Page 163. *"The forest primeval! . . . the mountain fastness."* Zane Grey, *Zane Grey's Adventures in Fishing.* Ed Zern, ed. New York: Harper & Bros., 1952, pp. 11-29.

Page 163. *A few months later . . . work during the night.* The Seattle Times, October 23, 1985, p. D2.

Page 165. *It was a rain-on-snow event . . . acres of the lake.* The Weekly, Seattle, March 2-8, 1983, p. 46.

Page 165. *. . . settling for $3 million out of court.* Paul Luvera, interview with the author, February 2, 1988.

Pages 165-166. *A report . . . "may be at critical levels."* Executive Summary: *Interagency Environmental Monitoring Program for Deer Creek Basin.* Unpublished paper, 1985.

Page 167. *In March 1983, Jeff Sirmon . . . in support of timber sales.* Jim Blomquist, Bill Arthur, and Barbara Boyle, *Forest Service Budget: Restoring the Balance.* Sierra Club, TWS, and NAS, March 1985, pp. 2-3.

Page 167. *Judge William H. Orrick, Jr. . . . "existence of fish to be taken" . . .* Fay G. Cohen, *Treaties on Trial.* Seattle: UW Press, 1986, p. 142.

Page 168. *Waldo outlined . . . negotiate with the tribes.* James C. Waldo, *U.S. v. Washington, Phase II: Analysis and Recommendations.* Prepared for the Northwest Water Resources Committee, September 1981.

Page 169. *On Puget Sound . . . Snoqualmie National Forest.* USFS, PNWReg, *Public Summary, MBS Plan,* p. 8.

Page 169. *The "Timber/Fish/Wildlife" . . . Bledsoe jumped at the chance. Puget Sound Business Journal,* January 5, 1987, p. 10.

Page 170. *The study by Cederholm . . . and very young fish.* C.J. Cederholm and E.O. Salo, *The Effects of Logging Road Landslide Siltation on the Salmon and Trout Spawning Gravels of Stequaleho Creek and the Clearwater River Basin, Jefferson County, Washington, 1972-1978. Final Report---Part III.* UW College of Fisheries, Fisheries Resource Institute, September 1979.

Page 171. *The TFW agreement . . . defined by mutual consent. Timber/Fish/Wildlife: A Better Future in Our Woods and Streams. Final Report,* February 17, 1987, in WDNR, *DEIS: Proposed Forest Practices Rules and Regulations. Appendixes.* May, 1987.

Page 172. *In 1932 . . . three-to-one margin.* Wells Associates, Inc., *National Forest Trails: Neglected and Disappearing.* San Francisco: Sierra Club, 1985, pp. 2-3.

Page 172. *Only in the early 1970s . . . (peak in 1944).* Wells Associates, Inc., *Protecting Trails in the National Forest Planning Process.* San Francisco: National Trails Coalition, c/o Sierra Club, 1986, p. 5.

Pages 172-173. *Between 1975 and . . . goals by 44 percent.* National Trails Coalition, *Toward a Balanced Budget: Conservationists' Proposals for the Forest Service Fiscal Year 1987 Budget.* May 1986.

Page 173. *. . . $50 million more than the administration requested.* Sierra Club *National News Review,* January 30, 1987.

Page 173. *. . . anonymous phone calls . . . wilderness designations.* The Oregonian, November 25, 1984.

Page 174. *. . . $6 billion annually. . . .* Visitors from out of state account for roughly half of this figure. Advance Washington, *"Tourism: It produces more than $3 billion a year in Washington state!"* and Rebecca Johnson, "Oregon's Forest-Related Recreation and the State's Economy," in *Assessment of Oregon's Forests,* p. 148.

Page 174. *In a survey of Washington tourists . . .* Gilmore Research Group, *1984 Washington State Conversion Study.* Prepared for WDT&ED, February 1984.

Page 174. *. . . Recreational Equipment . . . employs 1,400.* Also *. . . more than $800 million annually . . .* Ira Spring, "Trail Fact Sheet." Undated.

Page 174. *. . . nearly one-third of Seattle residents . . . Pacific (The Seattle Times* Sunday supplement), March 6, 1988, p. 6.

Pages 174-175. *Californians . . . the state's diversity.* Cole & Weber, *California Visitor and Prospect Qualitative Research.* Prepared for WDT&ED, Tourism Development Division, January 1986.

Page 175. *Olympic National Park . . . three million visitors . . .* In some years,

Olympic Park has ranked fifth in public use, in other years fourth. Rhoda Hansen, interview with author, May 8, 1989.

Page 175. . . . *President Franklin Roosevelt . . . "roasting in hell!"* Morgan, *The Last Wilderness*, p. 185.

Page 176. *A University of Delaware study . . . Seattle Post-Intelligencer*, April 16, 1987, p. C3. Study was conducted by Roger Ulrich and Robert Simon.

Pages 176-177. *"I was stunned . . . wasteland you've created." The Oregonian*, September 12, 1987.

Page 177. *"After driving . . . return home with." The Seattle Times*, September 30, 1988.

Page 180. *In six national forests . . . recreation permits than timber.* O'Toole, *Reforming the Forest Service*, esp. pp. 125-140, 226.

Page 182. *By the late 1990s . . . four-to-one margin.* USFS, PNWReg, *Public Summary, MBS Plan*, p. 39; *DEIS, MBS Plan*, pp. II-98-99; and *DEIS Appendices, MBS Plan*, p. B-142.

Page 182. *"Every worker . . . will be reduced."* W. Ed Whitelaw and Ernest G. Niemi, "The Greening of the Economy," *Forest Watch*, April 1989, p. 11.

Page 182. *One lifelong Darrington logger . . . "just ruined."* Dewey Bryson interview.

CHAPTER 9

Page 185. *The agency expected . . . appeals of its forest plans.* Julia Wondolleck, "Obstacles and Opportunities for Resolving Forest Planning Disputes," *Forest Watch*, August 1988, p. 15.

Page 186. . . . *nearly half . . . in the roadless areas. MBS Plan*, p. 4-19, and *Appendices--DEIS, MBS Plan*, p. C-3.

Page 186. . . . *Doug MacWilliams . . . "another set of plans."* Address to Washington chapter of The Wildlife Society, Olympia, Washington, February 6, 1986.

Page 186. . . . *another 44,000 acres will have been cut down. . . . MBS Plan*, p. 4-34.

Page 189. *Roads . . . two-fifths of the roadless areas.* Also *One of the alternatives . . . 35 percent below the present level. Public Summary, MBS Plan*, pp. 13, 39-40.

Page 190. *"This encourages negative . . . other land users."* Wondolleck, p. 17.

Page 191. *Richard W. Behan . . . "proves invulnerable."* Behan, "A Plea for Constituency-Based Management," *American Forests*, July/August 1988, pp. 46-8.

Page 192. *Mount Baker . . . 220 million board feet to 191. Public Summary, MBS Plan*, p. 39.

Page 193. *In 1984, 27 . . . two-thirds of their timber. Public Summary, MBS Plan*, p. 8.

Page 193. . . . *25,000 jobs in Puget Sound communities.* Also, *Those mills . . . 200 jobs and $10 million.* Brian Long, *Economic Impact Analysis of the USFS Proposed Plan for the Mt. Baker-Snoqualmie NF on King, Pierce, Skagit, Snohomish, and Whatcom Counties* (report to the Working Forest Alliance, March 1988), and The Working Forest Alliance, *The Mount Baker-Snoqualmie Draft Forest Plan: The Working Forest Alliance Viewpoint* (1988).

Page 193. . . . *sold an estimated 260 million . . . cut a whopping 295 million . . .* Ron DeHart, interview with author, February 9, 1988.

Page 194. *"[T]he tree farms . . . private tree farms."* Sampson, "A New Charter for the National Forests?" *American Forests*, May/June 1988, p. 18.

Page 196. *The national forest's figures . . . diameter at breast height.* Also *the average age . . . to 259 years. MBS Plan*, p. 4-34, and *DEIS, MBS Plan*, p. III-27.

Page 197. *Three national forests . . . Definition Task Group.* TWS, *End of the Ancient Forests: Special Report on National Forest Plans in the Pacific Northwest.* Washington, D.C.: June 1988, pp. A-1, A-2.

Pages 197-198. *The Wilderness Society . . . information on logs. Growth in the*

Pacific Northwest: A Status Report. Washington, D.C.: TWS, November 1988.

Pages 198-200. *After removing . . . would remain untouched. MBS Plan*, p. 4-34.

Page 200. *. . . Olympic National Forest . . . escape the chain saw. The Seattle Times*, November 17, 1986, p. B2, and November 19, 1986, p. H3.

Page 200. *Oregon's Willamette . . . slated to be cut.* USFS, PNWReg, *Proposed Land and Resource Management Plan, Willamette NF*, 1987, pp. II-5, IV-8, IV-9.

Page 200. *By 1981 . . . would remain uncut.* USFS, PNWReg, *Reader's Guide, DEIS, Proposed Land and Resource Management Plan, Gifford Pinchot NF*, pp. 11, 89.

Page 200. *Staff officers . . . not to save old-growth forests.* Gifford Pinchot NF, *Forest Planning Report*, December 1988.

Page 201. *So much old growth . . . "maximum habitat potential." Public Summary, MBS Plan*, p. 40.

Page 201. *. . . spotted owl pairs . . . ranged over 3,800 acres.* Harriet Allen, presentation to USFS chief and staff, Washington, D.C., July 27, 1987.

Page 202. *By the end of 1988 . . . threatened by logging.* WDW, appeal of USFS Record of Decision on *FSEIS*, January 23, 1989.

Page 202. *The Mount Baker-Snoqualmie's . . . some existing wilderness.* Trails 2000, "Trail Mileage Summary" (1988) and Washington Trails Association, "A Proposal for Backcountry Areas" (1988).

Page 202. *. . . less than one-sixth of projected future demand. DEIS, MBS Plan*, p. II-99.

Page 203. *The staff scenario . . . 35-percent cutback. Public Summary, MBS Plan*, p. 39.

Page 203. *Environmentalists . . . took the agency to court . . .* U.S. District Court Judge William Dwyer restricted Forest Service timber sales in spotted owl habitat, and the U.S. Ninth Circuit Court of Appeals issued an emergency injunction limiting BLM timber sales. With the federal timber program in chaos, Oregon's congressional delegation proposed a stopgap plan to supply logs to independent mills. That compromise" as it was billed, would have stripped the courts of authority to review federal timber sales and would have allowed *more* logging in fiscal year 1990 than proposed in the Forest Service's draft forest plans. In July 1989, Washington Senator Brock Adams and Oregon Senator Mark Hatfield gained the support of the Senate Appropriations Committee for the plan. Serious discussions of a long-term solution had yet to begin.

Page 204. *"Quite frankly . . . kick a little ass."* Jay D. Hair, address to Ancient Forests Conference, Portland, September 24, 1988.

CHAPTER 10

Pages 205-209. The Washington Wildlife Commission's hearing on the spotted owl was held January 15, 1988.

Page 208. *. . . U.S. Fish and Wildlife Service . . . was not endangered.* Rolf L. Wallenstrom, "Twelve-Month Petition Finding--Northern Spotted Owl," memorandum to director, USFWS, December 17, 1987.

Page 209. *In the fall of 1971 . . . goshawk in jeopardy.* E. Charles Meslow, interview with the author, January 6, 1988.

Page 210. *. . . 95 percent of their sightings . . . at least a century old.* Eric D. Forsman, "The Spotted Owl: Literature Review," Appendix C, *FSEIS* Vol. 2. Subsequent references to Forsman's research are also drawn from this literature review.

Page 210. *Two-thirds . . . Bureau of Land Management.* Also *Nonfederal owners . . .*

in the same period. FSEIS Vol. 1, pp. III-31, III-33.

Pages 210-211. *"Management of old-growth timber . . . supported the need."* Philip L. Lee, "History and Current Status of Spotted Owl (*Strix Occidentalis*) Habitat Management in the Pacific Northwest Region, USDA, Forest Service," in *Ecology and Management of the Spotted Owl in the Pacific Northwest.* USFS, PNWF&RExSta, GTR PNW-185, September 1985, p. 6.

Page 211. *In February 1977 . . . forests in Oregon and Washington.* Management history is drawn from *FSEIS* Vol. 1, pp. I-8 to I-16.

Page 212. *Douglas W. MacCleery . . . protection of the owl.* McCleery, letter to R. Max Peterson, March 8, 1985, in *FSEIS* Vol. 2, Appendix E.

Page 212. *The new Forest Service plan . . . spotted owl's range.* USFS, PNWReg, *DSEIS,* Vol 1.

Page 213. *Bruce Engel . . . owls on second growth? The Oregonian,* November 3, 1987.

Page 213. *Population projections . . . additional forest habitat.* USFS, *DSEIS,* Vol. 1., 4-25 to 4-33.

Page 214. *. . . owl pairs using 4,200 acres of old growth.* Harriet Allen, address at Earth Day, UW, April 1986.

Page 214. *(A more definitive figure . . . 3,800 acres.)* Harriet Allen, presentation to USFS chief and staff, Washington, D.C., July 27, 1987.

Page 214. *. . . National Audubon Society . . . "Oregon, and Washington."* Dawson *et al., Report of the Advisory Panel on the Spotted Owl,* especially pp. 31-4.

Page 214. *. . . fires that destroyed . . . old-growth forest.* USFWS, Region 1, *The Northern Spotted Owl: Status Review.* Portland: December 14, 1987, p. 22.

Pages 214-215. *The spotted owl subcommittee . . . "northern spotted owl."* Spotted Owl Subcommittee of Oregon-Washington Interagency Wildlife Committee, comments on *DSEIS,* November 3, 1986.

Page 215. *Considerably* less *owl habitat would be protected . .* The *DSEIS* proposed to protect between 313,839 and 690,446 acres of owl habitat (Vol. 1, p. 2-20), while the final plan modified slightly from the *FSEIS* would protect 374,000 acres (F. Dale Robertson, *Record of Decision, FSEIS,* December 8, 1988).

Page 215. *The Washington Wildlife Department . . . Curt Smitch.* Letter to F. Dale Robertson, September 29, 1988. The Forest Service, by contrast, offered new population projections less bleak under the new plan than under the earlier plan that would have protected nearly twice as much habitat. When ecologist Peter Morrison applied the Forest Service's definition of suitable owl habitat to Forest Service field data, he found less than half the habitat reported by the agency. One reason for the discrepancy, he discovered by talking to Forest Service biologists, was that the agency had adopted its definition *after* determining how much habitat existed.

Page 215. *A member of the condor . . . disappearing so rapidly.* Michael Soulé, letter to Mark L. Shaffer, November 1, 1987.

Pages 215-216. *. . . Forest Service Chief . . . information then available.* Robertson, *Record of Decision.*

Page 217. *Agency biologists . . . draft SEIS.* James F. Gore, interview with the author, February 9, 1988.

Page 217. *. . . "more rapidly than habitat destruction."* Mark L. Shaffer, *Assessment of Population Viability Projections for the Northern Spotted Owl* (Strix occidentalis caurina*),* November 19, 1987, p. 20.

Page 217. *In December 1987 . . . Forest Service's efforts.* Dunkle and Robertson, "Interagency Agreement," December 1, 1987.

Page 217. *The General Accounting Office . . . "to list the owl." Seattle Post-Intelli-*

gencer, February 21, 1989, p. B3, and February 23, 1989, p. A10.

Page 217. *Mark Boyce . . . "something I wrote."* Boyce, letter to Wallenstrom, February 18, 1988.

Pages 218-220. Memoranda from Gerald D. Patton, Steve Robinson, and Charles M. Luscher were written in September 1988. The Interior Department's official comments, signed by Bruce Blanchard, were dated October 18, 1988.

Pages 220-221. *During that two-year survey . . . "spotted owl population."* WDW, *A Survey of Spotted Owls in Western Olympic National Park*, September 1987.

Page 221. *DNR biologist Deborah Lindley . . . "owls on the Peninsula."* Letter to John Calhoun, May 20, 1987.

Page 222. *. . . The Seattle Times . . . "school kids." The Seattle Times*, December 15, 1987, p. A10.

Page 222. *"I get upset . . . cutting the trees too fast." The Seattle Times*, March 2, 1988, p. E2.

Page 222. *The Commission on Old Growth Alternatives . . .* In July 1989 the Board of Natural Resources accepted the commission's creative recommendations. Fifteen thousand acres of natural forest--the least-fragmented spotted owl habitat left--would remain in its natural state for at least 15 years. Meanwhile, the state would undertake selective logging and natural regeneration on a 260,000-acre "experimental forest" adjacent to the old growth.

Page 224. *. . . Todd True . . . "real habitat destruction."* Oral argument in U.S. District Court, Seattle, June 16, 1988.

Page 225. *Bruce Blanchard . . . "protection areas."* Letter to James F. Torrence, November 19, 1986.

Page 225. *Perhaps the most . . . "alternative possible."* Spotted Owl Subcommittee, Oregon-Washington Interagency Wildlife Committee, *Interagency Management Guidelines for the Northern Spotted Owl in Washington, Oregon, and California*, April 15, 1988.

Page 226. *"If it is an endangered . . . a smaller area." Seattle Post-Intelligencer*, March 16, 1989, p. B3.

Page 226. *Cason . . . Endangered Species Act.* Rolf Wallenstrom named Cason as one of the Interior Department executives who pressured him not to list the spotted owl as threatened. *Seattle Times*, April 11, 1989, p. A3.

Page 226. *less than a 1.5-percent drop.* National forests account for one-third of Oregon and Washington's timber supply. Thus, reducing the timber base of the affected national forests by 4 percent (Robertson, *Record of Decision*) translates into less than a 1.5-percent drop in the number of logs on the market. Since some national forests are not affected, the impact is even less than the numbers suggest.

Page 227. *. . . 60 million to 90 million board feet of timber sales . . .* In order to comply with the latest spotted owl protection plan, the Olympic National Forest officials expected the allowable sale quantity to drop from 210 million board feet to somewhere between 120 million and 153 million. Bob Burns, interview with the author, March 9, 1989.

Page 227. *. . . one billion board feet . . . from Olympic Peninsula ports . . . Warren, Production, Prices, Employment and Trade,* p. 28. The peninsula ports are Port Angeles and Grays Harbor (Aberdeen and Hoquiam).

CHAPTER 11

Earth First!'s blockades on Illabot Creek took place the week of August 18, 1987.

Page 233. . . . *Dave Foreman* . . . *"others should find them." The Seattle Times*, March 6, 1988, p. E1.

Pages 235-236. *Land would be bought* . . . *revenues could buy 28,000 acres.* Washington Public Lands Commissioner Brian Boyle in March 1989 proposed that the state issue $300 million in revenue bonds to buy approximately 300,000 acres of timberland to help prop up independent sawmills. The State Legislature, unable to come to grips quickly with the forest crisis, failed to act. Boyle's plan assumed that Scott Paper land could be bought for $1,000 per acre. Timber prices were rising fast, and Scott soon sold 194,000 acres to Crown Pacific Ltd. for close to $1,200 per acre. Expansion of the state lands would have eased an impending log shortage with little or no cost to taxpayers.

Page 236. *One-tenth* . . . *stocked with timber.* WDNR, *Washington Forest Productivity Study*, Phase III, Part II, p. 5.

Page 236. . . . *Philip L. Tedder* . . . *"price paid will be premium."* Address to conference, Small Woodlands Contribution to Oregon, Salem,   May 1 and 2, 1987.

Page 241. *"Ultimately* . . . *twenty-first century."* Jack Ward Thomas, address to symposium, Old-Growth Douglas-fir Forests, Portland, March 31, 1989.

Page 242. *"Oh, no* . . . *salvage it all."* O'Toole, *Reforming the Forest Service*, p. 98.

Pages 242. . . . *world's largest Douglas fir* . . . *another world title.* Olympic National Park, "Olympic: Official Map and Guide" (1988), and Herbert E. McLean, "Foray in Big Tree Country," *American Forests*, May/June 1988, pp. 46-9, 78.

Page 244. *Don't it always seem to go* . . . *A dollar and a half just to see 'em.* Joni Mitchell, "Big Yellow Taxi," 1974, Siquomb Publishing Corp. All rights reserved. Used by permission.

Page 245. *As little as 2 percent remains.* Jack Ward Thomas *et al.*, "Management and Conservation of Old-Growth Forests in the United States," *Wildlife Society Bulletin*, 1988, Vol. 16, No. 13, pp. 252-62. The authors state that 2 to 15 percent of the United States' original virgin forest, excluding Alaska's taiga, remains.

Index

KEITH ERVIN, a Seattle-based freelance writer, has covered environmental issues for a decade. His articles have appeared in *Sierra, Pacific Northwest, Washington, The Oregonian,* and *Seattle Weekly,* where he was a staff writer for three years. A native of Northern Virginia and a graduate of Oberlin College, he contributed to the *Point Reyes Light's* Pulitzer Prize-winning investigation of the Synanon cult in the late 1970s. This book is the result of four years of research on the struggle over the old-growth forests of the Pacific Northwest.